Marianne Deutsch

AF853347

COOLING TOWER TECHNOLOGY

Maintenance, Upgrading and Rebuilding

SECOND EDITION

COOLING TOWER TECHNOLOGY

Maintenance, Upgrading and Rebuilding

SECOND EDITION

Robert Burger

Published by

THE FAIRMONT PRESS, INC.
700 Indian Trail
Lilburn, GA 30247

Library of Congress Cataloging-in-Publication Data

Burger, Robert, 1921-
  Cooling tower technology: maintenance, upgrading & rebuilding.
    p. cm
  Bibliography: p.
  Includes index.
  1. Cooling towers. I. Title.

TJ563.B87 1989 621.1'97--dc20 89-45177
CIP

ISBN 0-88173-089-0

Cooling Tower Technology: Maintenance, Upgrading & Rebuilding
Second Edition
©1990 by The Fairmont Press, Inc. All rights reserved. No part of this publication may be reproduced or transmitted in any form or by any means, electronic or mechanical, including photocopy, recording, or any information storage and retrieval system, without permission in writing from the publisher.
Published by The Fairmont Press, Inc.
700 Indian Trail
Lilburn, GA 30247

Printed in the United States of America

10 9 8 7 6 5 4 3 2 1

ISBN 0-88173-089-0 FP
ISBN 0-13-173246-3 PH
While every effort is made to provide dependable information, the publisher, authors, and editors cannot be held responsible for any errors or omissions.
Distributed by Prentice-Hall, Inc.
A division of Simon & Schuster
Englewood Cliffs, NJ 07632

Prentice-Hall International (UK) Limited, London
Prentice-Hall of Australia Pty. Limited, Sydney
Prentice-Hall Canada Inc., Toronto
Prentice-Hall Hispanoamericana, S.A., Mexico
Prentice-Hall of India Private Limited, New Delhi
Prentice-Hall of Japan, Inc., Tokyo
Simon & Schuster Asia Pte. Ltd., Singapore
Editora Prentice-Hall do Brasil, Ltda., Rio de Janeiro

This book is dedicated to the women in my life:
present -- Ellamae, Donna, Jan, Debbie, Mary;
past -- Wanda; and future -- Genny, Lauren Rose,
Rachael Wanda...

CONTENTS

1 - The Common Denominator ... 1
2 - Basic Fundamentals ... 5
3 - Counterflow vs. Crossflow ... 11
4 - Water Distribution ... 21
5 - Drift Elimination ... 29
6 - Thermal Evaluation ... 35
7 - Air Management ... 65
8 - Structural Analysis ... 87
9 - Natural Draft Cooling Towers ... 95
10 - Cooling Water Chemistry ... 103
11 - Siting ... 117
12 - Winter Operations ... 123
13 - Maintenance ... 133
14 - Inspection for Profit ... 149
15 - Fan Wreck Prevention ... 161
16 - Operating Concepts ... 165
17 - Bid Evaluation ... 175
18 - Field Inspection ... 185
19 - Redwood in Cooling Towers ... 189
20 - Rebuilding for Profit ... 199

Appendix 1: Glossary - The Language of Cooling Towers ... 231
Appendix 2: Metrification ... 247
Appendix 3: Psychometric Analysis ... 251

ABOUT THE AUTHOR

ROBERT BURGER is President of the Dallas-based mechanical contracting firm of Burger & Associates. Founded in 1958, the firm is now recognized as one of the leaders in the building, rebuilding and upgrading of cooling towers, having completed upgrading energy conservation projects and new tower erections across the U.S., Canada and Mexico, and overseas in Singapore, The Philippines, Brazil, Puerto Rico and Malaysia. The firm has recently contracted to perform work for a subsidiary of the Indian government.

Mr. Burger has also written widely on the topic of cooling towers, including over 30 articles in magazines such as *National Engineer, Power, Chemical Engineering, Plant Engineering, Heating/Piping/Air Conditioning, Consulting Engineer, ASHRAE Journal, Journal of the Cooling Tower Institute, PT/Procestechniek* (The Netherlands), and many others. He also authored the chapter on cooling towers for the Prentice-Hall *Operations and Maintenance of Equipment Manual* and for Ullman's *Encyclopedia of Industrial Chemistry*, in addition to this book, which represents the only comprehensive, plain language reference covering all aspects of cooling tower technology.

He lectures extensively both in the U.S. and overseas, having appeared before a wide range of organizations, such as the Petro-Chemical Institute, the American Institute of Chemical Engineers, the U.S. Environmental Protection Agency, the American Society for Heating, Refrigeration, and Air Conditioning Engineers, the National Association of Power Engineers, the Refrigeration Engineers and Technicians Association, the Cooling Tower Institute, Exxon Refineries, Eastman Kodak and NASA. He conducts several two-day seminars each year for the Association of Energy Engineers, as well as seminars in alternate years in Kuwait and The Netherlands, and was seminar leader for the McGraw Hill cooling tower program.

As a long time member of the Cooling Tower Institute, Mr. Burger was founding editor of the *Journal of the Cooling Tower Institute*, and has served the Institute's Engineering Standards Committee, Performance Technology Committee, and Board of Directors as Publications Chairman and Public Relations Committee Chairman.

Mr. Burger received his A.E.M.E. from Cornell University in 1943. He is a member of the Cornell Society of Engineers, Cooling Tower Institute, National Association of Power Engineers, Aiche, and Flying Engineers. For his educational efforts and interest in the National Association of Power Engineers, the Past President's Selection Committee of that organization named Mr. Burger "Engineer of the Year" for 1989.

ACKNOWLEDGEMENTS

Acknowledgement, with sincere thanks, is made to the following who shared their expertise; Publications, for permission to reprint copyrighted excerpts of the author's articles; and Organizations who permitted the use of their technical materials.

AIPE JOURNAL:..."Cooling Towers, The Energy Savings Machine".

ASHRAE:...Psychrometric Charts, Pages 20-22, ASHRAE Handbook, 1988 Equipment Volume.

CHEMICAL ENGINEERING, MCGRAW HILL PUBLICATIONS:... "Cooling Tower Asbestos Abatement", "Retrofitting Cuts Cooling Tower Costs".

CHEMICAL ENGINEERING PROGRESS:..."Drift Eliminators Save Energy", "Basic Cooling Tower Technology".

CONSULTING ENGINEER:..."Siting of Cooling Towers".

CORNELL ENGINEER:..."Cooling Tower, Money Making Energy Machine".

DEPARTMENT OF ENERGY:...Natural Draft Cooling Tower data.

FAVRA INTERNATIONAL:...Metric conversion tables.

GORDIAN ASSOCIATES:...Report to Department of Energy, "Relative Economics and Total Energy Requirements of Natural Draft and Mechanical Draft Cooling Towers".

HEATING/PIPING/AIR CONDITIONING:..."Cooling Tower Technology".

HUDSON PRODUCTS CORPORATION:...Fan Rating Procedures.

INDUSTRIAL WATER ENGINEERING:..."Basic Cooling Tower Technology", "Field Inspection and Certified Testing".

THE MOORE COMPANY:...Fan Selection Charts.

NATIONAL ENGINEER:..."Language of Cooling Towers", "Kilowatt Recovery Via Velocity Regain", "Maintenance and Repair Makes Money", Upgrading For Profits", "Modernizing Cooling Towers".

NATIONAL GEOGRAPHIC SOCIETY:...Lenart Nilsson Boehringer Ingleheim, Zentrale Gnb H, Macrophage Photo Micrographs.

PLANT ENGINEERING:..."Inspect for Profit", "Glossary of Cooling Towers", "Counterflow vs. Crossflow".

POWER:..."Cooling Tower Certification", "Retro-fit to conserve Energy".

POWER ENGINEERING:..."Performance Monitoring Enhances Power Plant Operations".
DR. SIDNEY SUSSMAN:..."Fundamentals of Cooling Tower Technology" presented by Dr. Sussman to C.T.I. Annual Meeting, Houston.

INTRODUCTION

A great majority of cooling towers operating today, even though some are newly installed, have been engineered with techniques over 30 years old. Today's state of the art can be utilized to retrofit practically all towers and upgrade their capability for producing colder water or cooling greater volumes of circulating water in the same tower.

A recent government agency survey stated, "Inadequate thermal performance of cooling towers surveyed is costing electric utilities more than $25 million per year in lost revenue and higher fuel costs. Of the plants TVA surveyed, sixty-five per cent had towers that either failed to meet design specifications or had been the cause of decreased plant efficiency or capacity. This is mainly due to higher back pressures on the turbines which decreased the cycle efficiency."

In process plants, power generating stations, chemical plants, and refrigeration systems that depend upon the dissipation of waste heat, it behooves the owner and operator to investigate the possibility of upgrading the existing cooling tower rather than installing another O.E.M. unit which may or may not produce the necessary colder water. It should be understood that colder water can save energy and create an operating profit. Cooling towers are hidden bonanzas for energy conservation and dollar savings when properly engineered and maintained. In many cases, the limiting factor of production is the quality of cold water coming off the tower.

Most towers designed over 15 years ago were inefficiently engineered due to cheap power and the "low bidder" syndrome. Operating energy costs were ignored and purchasing criteria was to award the contract to the lowest bidder.

It would be prudent for the engineer with responsibility for the efficient operation of refrigeration systems or any other process dependent upon the dissipation of waste heat to have a professional inspection to analyze the energy savings potential of the installation. In these days of high energy costs, the savings accrued from a well engineered and retrofitted cooling tower will make a significant impact on a company's profit and loss statement.

This book investigates basic thermal transfer on a practical basis and looks at the internal elements of typical towers, delineates their functions and shows how to upgrade them in the real world for energy savings and profitability of operation.

- Robert Burger

CHAPTER 1

THE COMMON DENOMINATOR

FIG. 1. Chemical plants

Chemical plants (Fig. 1) throughout the country yield an unimaginable number and variety of products our civilization utilizes. In all of our activities, day and night, we rely upon them, and in these chemical operations, there is one common denominator.

Without electric generating plants (Fig. 2), the fabric of our society would collapse, whether they be a small 1 megawatt (1-MW) installation for a small apartment house complex burning fossil fuel, or a 1,000-MW Nuclear steam-electric facility for a large public utility, there is one common denominator.

The New Orleans Superdome (Fig. 3) air conditioning/refrigeration system is rated at 10,000 tons circulating 32,000 gallons per minute of water, while a small town Shoe Store cools it's premises with a 10 ton air conditioning central system circulating 30 GPM. There is one common denominator.

This common denominator cannot be ignored, it is WASTE HEAT!!

In the above examples, waste heat is absorbed by water having a temperature colder than the process. This hot water must now be either discharged into a body of water or cooled and recycled. The alternative similar to all systems is that the hot water temperature is lowered in a device called a cooling tower whereby the waste heat is rejected into the atmosphere, and the colder water is then recirculated throughout the system thereby conserving our water resources.

The problem is enormous since industry uses approximately 500 billion gallons of water a day, which if not recirculated would seriously deplete the available water supply. The unwanted heat from the water must be restored to the environment without creating ecological havoc or adding un-necessary expenses to the inflationary spiral.

The development of cooling towers solved the problem of possible ecological damage of returning hotter water to the environment and enables industry to

recycle the greater proportion of the water it requires at reasonable capital investment costs.

Operating procedures, costs, and upgrading potential for greater efficiencies must be constantly investigated to provide economic returns for the installation. To intelligently achieve this, the fundamental principals of how the cooling tower performs, identification and understanding of the various elements are necessary to be able to upgrade them. A comprehensive maintenance program must be instituted to obtain maximum utilization of the equipment.

FIG. 3. The New Orleans Superdome

CHAPTER 2

BASIC FUNDAMENTALS

Fig. 1: Evaporative condensation, the main principle of cooling tower operation, creates a loss of water volume in the form of vapor discharge.

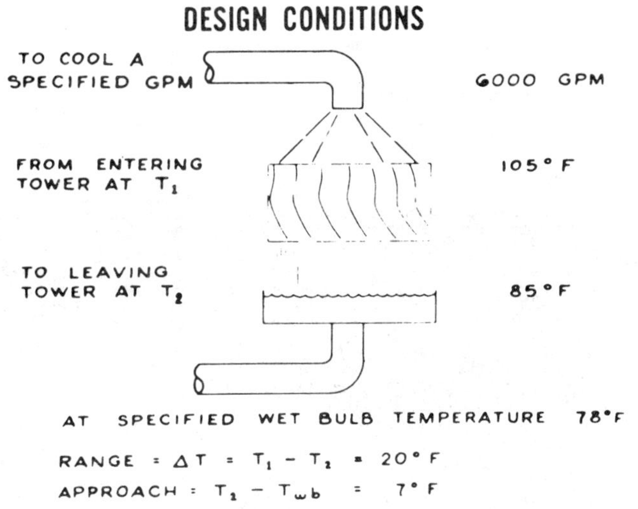

Fig. 1. Design conditions describe the duty and work expected from a particular cooling tower.

Operational Responsibility

All too many engineers pay a minimum of attention to the cooling tower, being aware only that it is that "Box in the back where we send the hot water and it comes back cold".

Many managers are not even aware, or do not consider the role of the cooling tower in their company's profit and loss picture. It is the engineers' function to employ the equipment in as efficient a manner as possible to produce money and profits for Management.

The major area of profit to analyze is the utilization of the cold water. In the refrigeration/air-conditioning system, the colder the water, the less energy is required to power the equipment. In chemical product manufacturing plants, the colder the water, the more efficient the condensation and the greater the volume of saleable condensate product yielded for the plant at lower cost. In power plants, colder water reduces the electrical generation penalty (the cost of electricity used in the process of producing saleable or usable power).

The second area of profit to consider is the electrical and/or steam dollar expenditure which is purchased to power the systems. The cooling tower is a high energy utilization machine as well as a mass and heat transfer device.

After the cooling tower has been erected, proper operation and maintenance is required to yield the design conditions of cooling. By not maintaining towers adequately, they can lose appreciable performance as soon as they are started up.

This paper on cooling towers is aimed at one basic end result: — . . . COLD WATER . . .

Design Conditions

The criterion of cooling tower performance is outlined in the Design Conditions, Fig. 1, specified when the cooling tower is purchased and/or rebuilt to mean the cooling of a specific quantity of circulating water from entering the tower at a particular temperature and leaving at a definite value. The difference betwen these two temperatures is called "The Range" or "Delta T". This cooling of the specific amount of water is also to be performed at a wet bulb temperature that is noted in the specifications. The difference between the cold discharged water and wet bulb is called "THE APPROACH to the wet bulb".

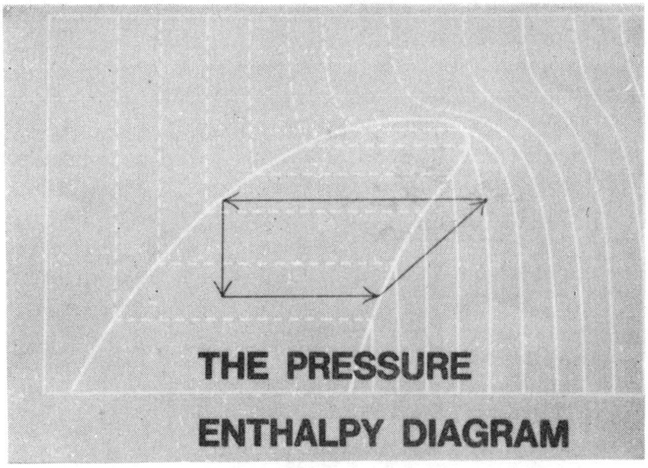

Fig. 2. Enthalpy pressure diagram relates cold water to energy required for refrigerant utilization.

A 1° F (0.6° C) colder water returned to the compressors and condensors in air-conditioning and refrigeration equipment calculates, by use of Enthalpy Charts, Fig. 2, to a 3% savings in electrical energy input to these machines. Therefore a little more than 3° F (1.7° C) colder water off the tower can save 10% of electrical energy and resulting charges thereof at any given time. A 2000 Ton (1815 metric ton) refrigeration system circulating 6,000 GPM (22,700 liters/min) could use $350,000 in electrical power a year. A 10% savings of $35,000 obtained by sending colder water to the machinery will be significant factor towards rapid pay back for the upgrading of the tower.

Cooling Tower Fundamentals

Before proceeding further, we shall consider the fundamentals of cooling tower technology. The basic principle of cooling tower operation is that of evaporative condensation and exchange of sensible heat. The air and water mixture releases latent heat of vaporization. Water exposed to the atmosphere evaporates and as the water changes to vapor, heat is consumed which amounts to approximately 1,000 BTU per pound of water evaporated. The heat is taken from the water that remains by lowering its temperature.

However, there is a penalty involved, and that is loss of water which goes up the cooling tower and is discharged into the atmosphere as hot moist water vapor. Under normal operating conditions this amounts to approximately one and two tenths percent (1.2%) for each 10° F (5.5° C) of cooling range. Sensible heat that changes temperature is also responsible for part of the cooling tower's operation. When water is warmer than the air, there is a tendancy for the air to cool the water. The air then gets hotter as it gains the sensible heat of the water and the water is cooled as its sensible heat is transferred to the air.

Cooling due to the evaporative effect of the release of latent heat of vaporization amounts to approximately 75% while 25% of the heat exchange in the cooling tower is sensible heat transfer. The cooling tower, like any other device, process, or operation does not escape the law of conservation of matter. A cooling tower is merely a machine that takes a mass of heat from one area and moves it to another area. In more technical terms, it is referred to as "The heat rejection solution of the chemical process" or "Correction of the heat penalty generation of compression equipment". The cooling tower is a machine that moves heat from point "A" to point "B" and ultimately discharges this heat into the atmosphere which Thermal Engineers euphemistically call a "Heat Sink".

Counterflow vs Crossflow

All cooling towers, except the small packaged specialty models, are custom-made and the manufacturer varies the shape, size, configuration and input to meet the particular set of thermal parameters required by the customer. Since they are available in two basic operational designs, Counterflow and Crossflow, it is useful to list a few terms and their definitions to aid in investigating these differences so that the nomenclature will be standard. The letters in the circles on the schematic diagrams, Fig. 3 and Fig. 4 refer to the elements that perform the same function and their location in the structure.

Upgradeable Elements

To upgrade the performance of an existing cooling tower, the three major areas to investigate for retro-fit are:

1. Wet Decking Fill
2. Water Distribution System
3. Drift Eliminators

Wet Decking Fill

Generally the most significant improvements can be made simply by changing the wet decking fill. This however is not done capriciously. The heat transfer must be investigated from a thermal engineering point of view in conjunction with the fill characteristics as determined by the manufacturers' performance curves which are developed very painstakingly by trial, error and experimentation which is expressed as KAV/L or heat transfer characteristics. It is quite obvious that this wood fill Fig. 5, is extremely inefficient compared to the cellular fill, plus the fact that the cellular fill has a lower

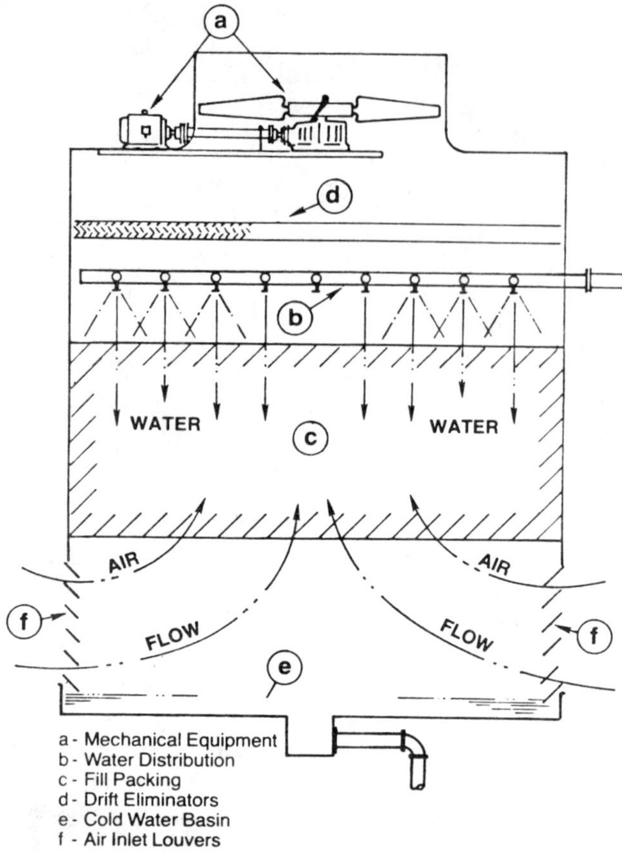

Fig. 3. Counterflow schematic

static pressure of operation which further enhances the heat transfer by more efficient utilization of the existing air. In wood slat splash bars, Fig. 6 and Fig. 7 droplets of water bounce from one layer of wood to the other and the rising air cools the outside of each sphere of the water droplets. Cellular fill takes the same droplet of water and spreads it out in a very thin molecular film where the air can now effect the entire surface of the film as in Fig. 8. Considerably more surface is then available to the flowing air for vaporization and sensible heat exchange to take place. The film pack contains more surface area than splash bars and since the design of cellular fill permits air to flow through it with less static pressure, it is extremely efficient compared to the old fashioned splash bar mixture system for the same perameters of cooling.

Water Distribution System

For maximum performance, the water distribution system must provide a uniform pattern over the fill for optimum air-water interface. The older type trough has a very uneven splash distribution based upon columns of water falling vertically and hitting cups which, when accurately placed, distribute the water throughout the fill. Over the years this delicate balance is destroyed as the tower deteriorates and the end result is a vertical column of water, in many places, dropping up to 4 feet (122 cm) through the fill before it is broken up, losing entire areas of efficiency. When a nozzle is clogged, it leaves a dry spot on the fill and the air follows the path of least resistance rushing up this dry spot, wasting energy and cooling potential. The main difference between counterflow towers as illustrated and crossflow is that the air in counterflow moves vertically through the fill while in the crossflow tower, air flows horizontally through the fill. In both cases, the water drops vertically through the fill

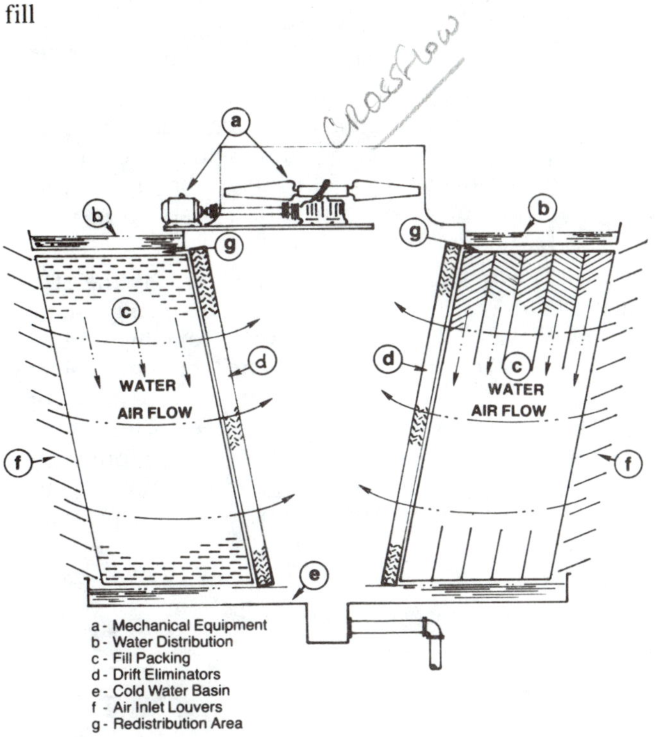

Fig. 4. Crossflow schematic

Fig. 5. Inefficient wood splash bar fill provides poor heat transfer

Fig. 6. Cellular fill replacement indicates 36″ doing the same work as 60″ more efficiently at lower static pressure.

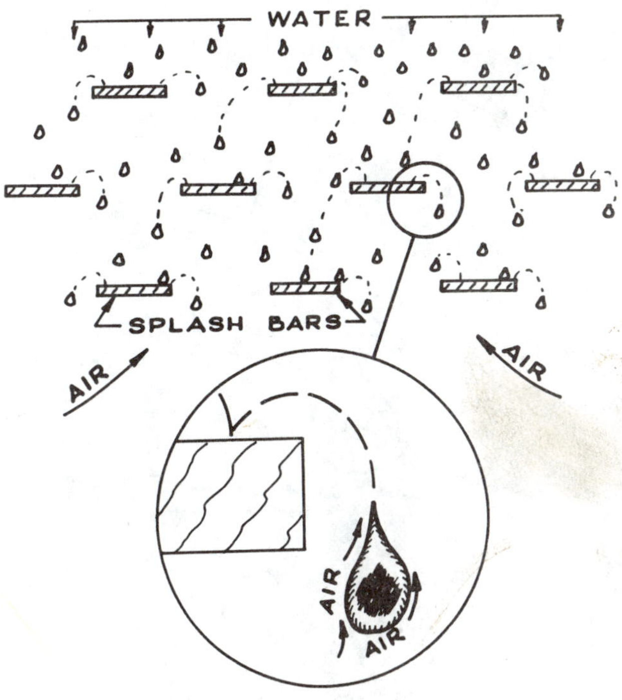

Fig. 6: Schematic of the cooling of droplets of water bouncing vertically from wood slat to wood slat. Only the exterior of the droplet is cooled.

Crossflow cooling towers also have a rebuilding capability to improve the performance by changing the wood splash bars to more efficient fill. Old fashioned stacked fill is rather inefficient compared to the new grid type splash bar packing consisting of corrosion proof fiberglass polyester grids for the higher capacity installation. A less expensive approach is to use metal grid hangers which are coated for corrosion protection. The installation of this type of fill is quite laborious, but is relatively inexpensive to install for the percentage of improvement that can be obtained compared to stacked fill.

The greatest improvement of performance in cross-flow fill, is obtained by changing the wood splash bars to cellular fill, Fig. 8, installing redistribution sections of predetermined levels and adjusting the flow rate throughout the hot water distribution basin to provide proper water loading through the fill.

Here again, wood splash bars are replaced by the efficient cellular fill sections thereby reducing static operating pressure through the tower, providing a more effective use of the input energy, and obtaining colder water from the tower. Whether it be crossflow or counterflow, **uniformity is the key to success.** Uniformity of water distribution, uniformity of static pressure, uniformity of fill configuration, and uniformity of air velocity, even though in a crossflow tower, the pressure differential will vary with the height.

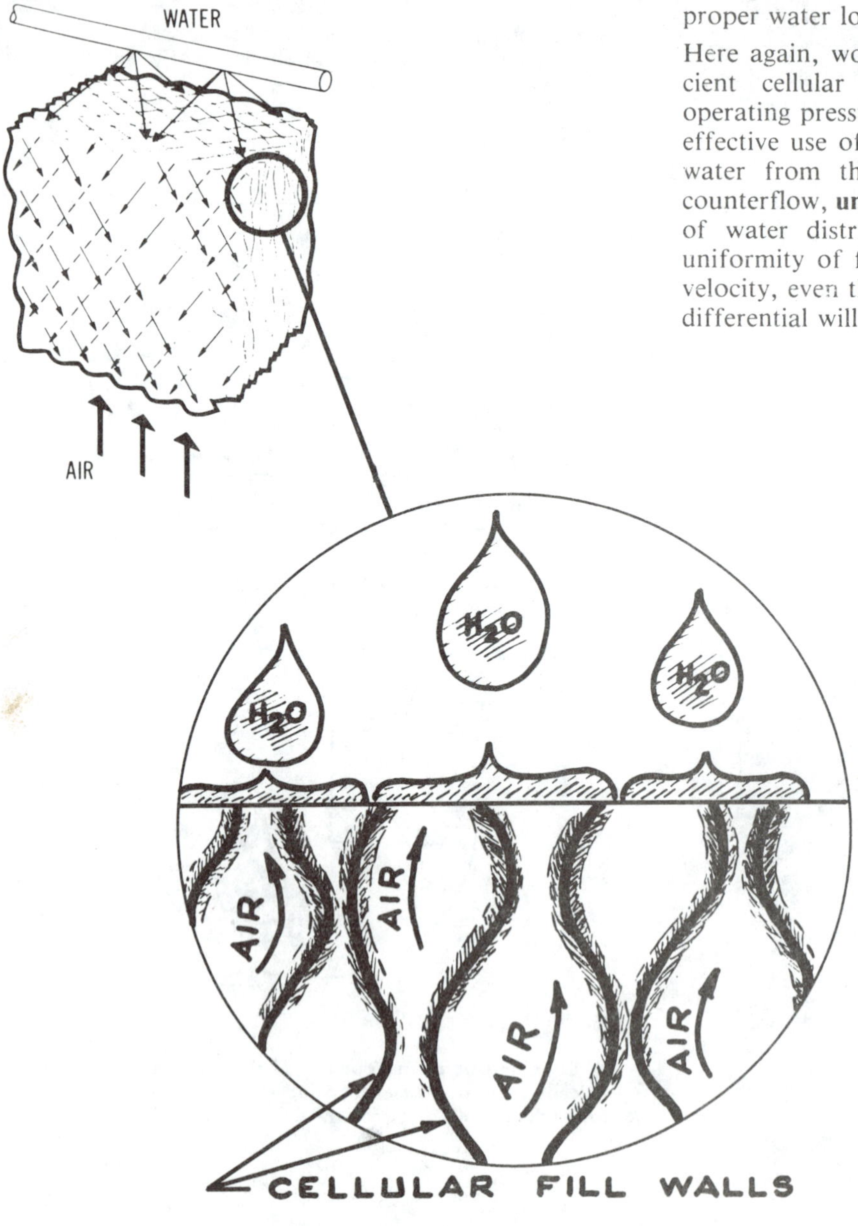

Fig. 8. Cellular (film) fill whereby the droplet of water is turned into a thin film providing more rapid and efficient air-water cooling interface.

CHAPTER 3

COUNTERFLOW VS. CROSSFLOW

Reduce pumping head horsepower, prevent air recirculation, control icing problems, and improve heat transfer capability by selecting

Counterflow vs Crossflow

During the 1960's, virtually all counterflow cooling tower manufacturers changed to crossflow designs, and by 1970 more than 90 percent of the field-erected, industrial, mechanical draft cooling towers in the United States were the crossflow type. This dramatic switch was largely attributed to the fact that cooling towers utilizing traditional wood slat splash bar fill, Fig. 1, had to be built increasingly larger to accommodate newer and greater heat rejection requirements in the plant. As cooling towers became larger, the counterflow design required more pumping horsepower and more heat-transfer surface area to cool water than similar capacity crossflow models.

The advent of high-efficiency, densely packed, cellular-fill heat-transfer material, Fig. 2, proved to be particularly beneficial to counterflow towers because it allowed the equipment to be built with much lower heights, lower pumping horsepower requirements, and lower initial costs than earlier versions and made the model competitive with the crossflow design.

Today's counterflow cooling towers offer a number of advantages over comparably sized crossflow models.

Pumping Head Horsepower—A large part of cooling tower operating costs results from the pumping head required to bring the circulat-

Fig. 1. Many older cooling towers have wood slat splash bar fill. Water droplets bounce off the layers of wood slats as air cools the outside of each drop. A large cross-sectional area is required for heat transfer, which made counterflow cooling towers much larger than equal capacity crossflow models.

Fig. 2. The advent of high-efficiency, densely packed, cellular-fill heat-transfer material proved to be particularly beneficial to counterflow towers because it allowed the equipment to be built with much lower heights, lower pumping horsepower requirements, and lower initial costs than earlier versions.

ing hot water to the top of the cooling tower. A counterflow tower, Fig. 3, requires 20 to 50 percent less pumping head than a crossflow tower, Fig. 4, of the same capacity. If a tower pumps 6000 gpm an extra 30 ft of head, the additional expense for electricity over the life of the equipment far exceeds the initial cost saving associated with a crossflow tower.

The counterflow tower requires an additional 4 to 6 ft of pumping head for the low-pressure water distribution spray system; however, the total pumping head requirement is still significantly less than it is for crossflow designs.

Recirculation—The fan draws air through the entire louver face of a crossflow tower; therefore, a pressure differential is created along the vertical plane of the structure. Air entering the upper sections is closer to the fan and travels through the tower at a higher velocity than air at lower levels. Bernoulli's theorem states that when the velocity of a fluid (air) is increased, its pressure is decreased; this effect produces negative pressure at the upper levels of the cooling tower and, under typical atmospheric wind conditions, sets up air recirculation, Fig. 5. Hot moist air with a high wet-bulb temperature is pulled back into the cooling tower, which greatly reduces its performance.

The air-intake louvers of a counterflow tower are near the base, a considerable distance from the fan discharge plume, so it is difficult for air to recirculate under normal conditions. Counterflow towers can have recirculation problems if the discharge plume hits a barrier adjacent to the fan outlet, which forces the hot moist air downward. This problem can be avoided by installing an elevated fan cylinder to discharge the air above the barrier.

Icing Control—One of the most effective methods of preventing icing problems in cooling towers is to

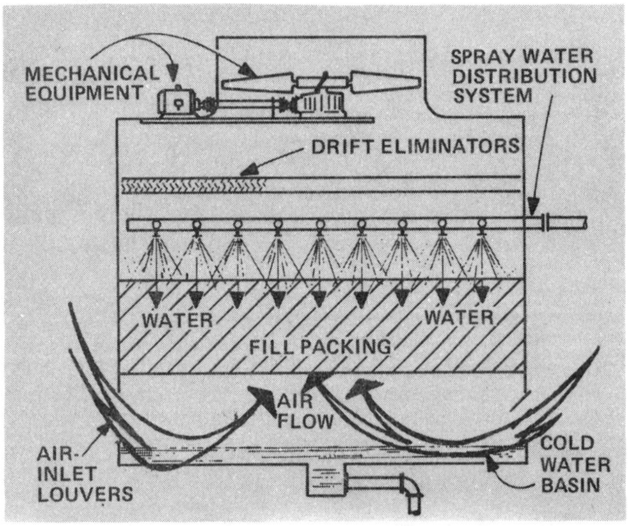

Fig. 3. Counterflow towers have air-inlet louvers near the base. Air enters horizontally from two or four sides, immediately turns upward, and encounters water falling from a low-pressure spray-distribution system. Water strikes splash bars of cellular fill to accelerate the heat-exchange process. Moisture-laden air rises through the distribution piping, past the drift eliminators, and into the plenum chamber at the top. The fan exhausts the air through the stack.

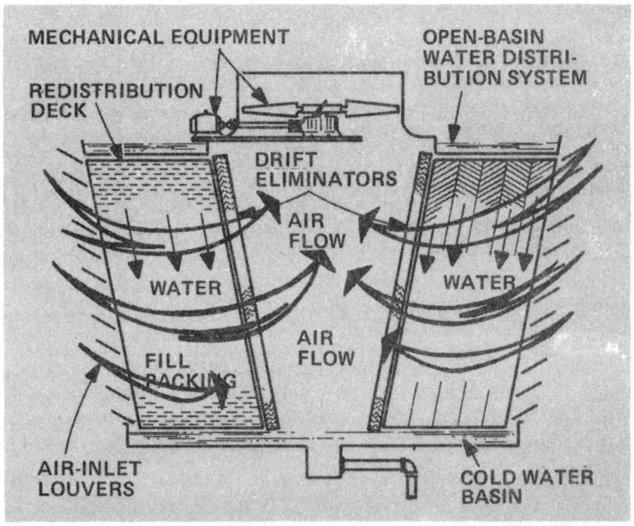

Fig. 4. Crossflow towers have air-inlet louvers located along the entire height of one or two sides. Air moves horizontally through the fill while water falls vertically, by gravity, from an open distribution basin on top of the tower. Air moves through the drift eliminators into a plenum chamber in the center of the tower. The fan then draws the air out of the tower.

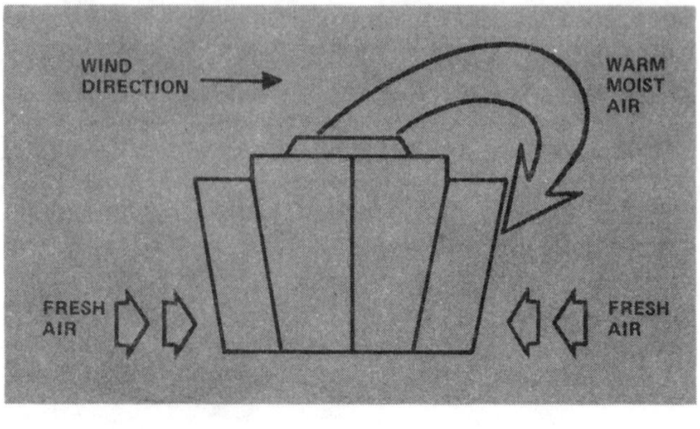

Fig. 5. The fan induces air through the entire louver face of a crossflow tower. Air entering the upper sections is closer to the fan and travels through the tower at a higher velocity than air at lower levels. Bernoulli's theorem states that when the velocity of a fluid (air) is increased, its pressure is decreased; this effect produces negative pressure at the upper levels of the cooling tower and, under typical atmospheric wind conditions, sets up air recirculation. Hot moist air with a high wet-bulb temperature is pulled back into the cooling tower, which reduces its performance capability.

Fig. 6. These performance curves are based on a 78 F wet-bulb ▶ temperature, 15 deg F range, 1.0 L/G, and 1.0 KaV/L (counterflow) or KaY/L (crossflow). The crossflow tower cools to a 12 deg F approach to the wet-bulb temperature, and the counterflow tower cools to a 10 deg F approach to the wet-bulb temperature; the closer the approach, the more desirable the tower.

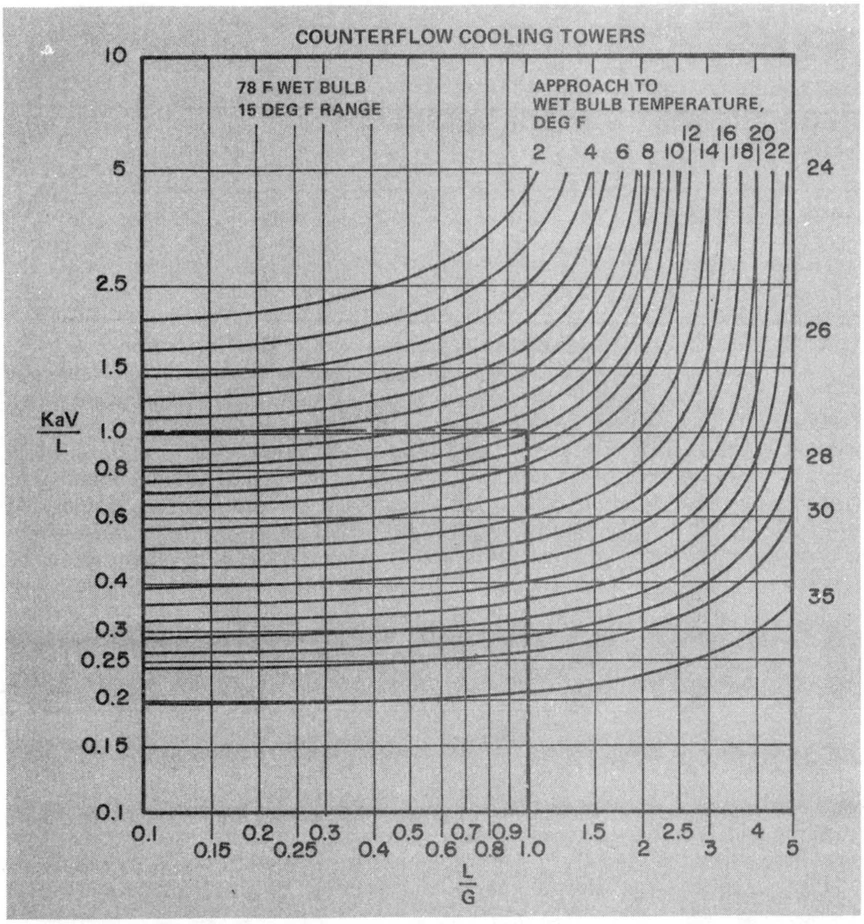

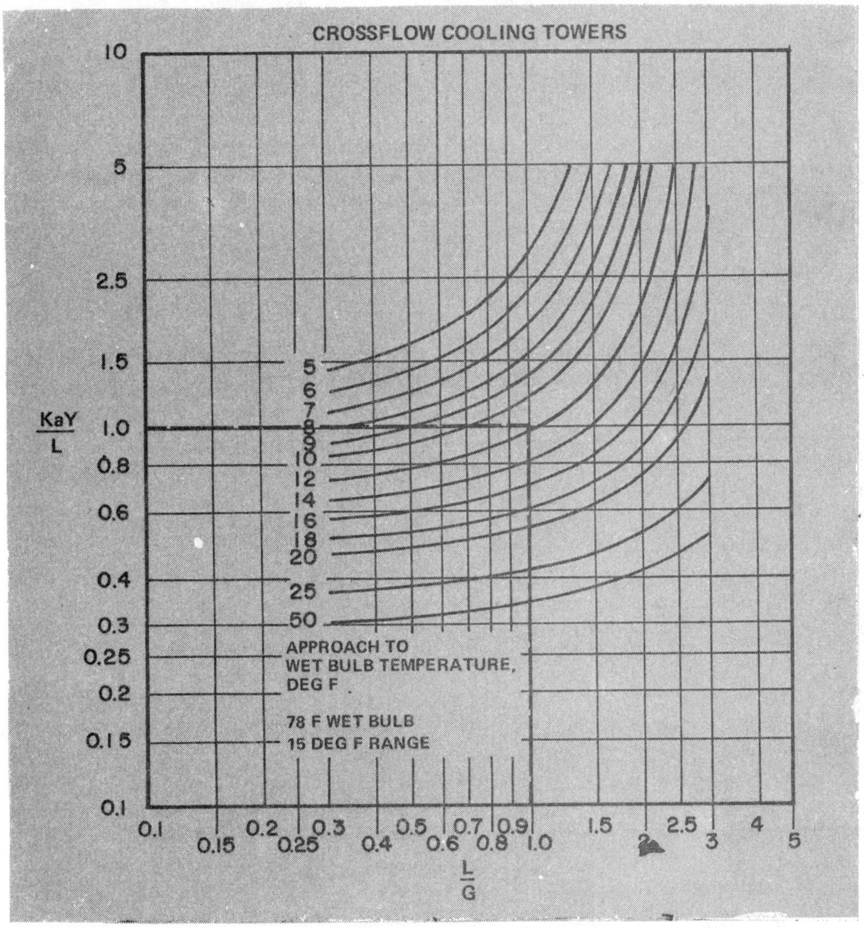

rotate the fans in reverse, which provides heat to melt the ice. This procedure is particularly effective in counterflow towers whose enclosed sides retain the heat.

Crossflow towers have louver openings all along the vertical sides and the heat rapidly dissipates into the atmosphere, especially at the lower levels where the most destructive icing forms.

Heat-Transfer Capability—Kelly's *Handbook of Crossflow Cooling Tower Performance* and the Cooling Tower Institute's *Cooling Tower Performance Curves* (for counterflow towers) reveal that a counterflow tower produces more cooling per unit of volume for less cost than a crossflow tower under the same design conditions. Figure 6 shows performance curves for crossflow and counterflow cooling towers. Both curves are based on a 78 F wet-bulb temperature, 15 deg F range, 1.0 L/G, and 1.0 KaV/L or KaY/L.

The crossflow tower cools to a 12 deg F approach to the wet-bulb temperature, and the counterflow tower cools to a 10 deg F approach to the wet-bulb temperature; the closer the approach, the more desirable the tower.

KaV/L or KaY/L, called the cooling tower characteristic, is a calculated value based on the equipment's design requirements; it has no relationship to operating conditions. Testing of the cooling tower after installation establishes a new characteristic curve of the actual operation.

L/G is the liquid-to-gas ratio; the liquid is water and the gas is air. This relationship is based on the required conditions of the plant's particular process. The L/G ratio conforms to the laws of thermal dynamics; notably, the enthalpy difference is the driving force of evaporative cooling within the tower. The design characteristic curve and the L/G ratio are furnished by the cooling tower manufacturer.

• **PUMPING HEAD** is an important evaluation item which can be calculated into horsepower, kilowatts, and consequent direct costs over the projected life of the tower. If is often neglected and it should be noted that some crossflow towers **(Fig. 7)** would require considerably more pumping head due to the design as compared to counterflow towers **(Fig. 8).** The table at the right indicates the dollar cost of additional horsepower pumping head. For example, the difference between a large crossflow and counterflow tower with 30′ of pumping head for 20,000 GPM of water will require an extra 190 HP. The table indicates that this will cost $ 55,244 if the particular tower is operated, like most power generating stations, 8760 hours per year. The operating cycle of your tower can be calculated and the chart used to determine the extra horsepower penalty cost of additional pumping head. This dollar figure can be subtracted from the lower pumping head tower for lower fan horsepower utilization, thereby indicating that a tower that might cost a higher price initially will, in the long run on a maintenance and power and utilization basis, be a much better buy for your company.

Figure 7

Extra cost of oversized fan or excessive pumping can be determined by chart.

$$BHP = \frac{G.P.M. \times PUMP\ HEAD}{3960 \times EFFICIENCY \times SP.GRAV.}$$

Motor Size HP	Full Load *KVa* @ 1800 RPM	Operating Cost Per Year (8760 hours) at Stated Cost per kWh ($)							
		.03	.04	.05	.06	.07	.08	.09	.10
10	10.14	$ 2,665	$ 3,553	$ 4,441	$ 5,330	$ 6,218	$ 7,106	$ 7,994	$ 8,883
15	15.67	4,118	5,491	6,863	8,236	9,609	10,982	12,354	13,727
20	20.06	5,271	7,029	8,786	10,544	12,301	14,058	15,815	17,573
25	24.93	6,551	8,735	10,919	13,103	15,287	17,471	19,655	21,839
30	28.58	7,511	10,014	12,518	15,022	17,525	20,029	22,532	25,036
40	38.32	10,070	13,427	16,784	20,141	23,498	26,855	30,211	33,568
50	47.10	12,378	16,504	20,630	24,756	28,882	33,008	37,134	41,260
60	57.21	15,035	20,046	25,058	30,070	35,081	40,093	45,104	50,116
75	71.10	18,685	24,913	31,142	37,370	43,599	49,827	56,055	62,284
100	91.11	23,944	31,925	39,906	47,887	55,869	63,850	71,831	79,812
125	118.47	31,134	41,512	51,890	62,268	72,646	83,024	93,402	103,780
150	140.57	36,942	49,256	61,570	73,884	86,198	98,511	110,825	123,139
200	188.65	49,577	66,103	82,629	99,154	115,680	132,206	148,732	165,257

Figure 8

To evaluate electric consumption, assume an example of two cooling towers, both pumping 20,000 GPM with power at 5¢ per KwH, one Counter-flow and the other Cross-flow configuration.

1. **FAN MOTORS**

   Counter-flow: 4 cells with 75 horsepower drivers = 300 HP.
   Cost = $82,629.00* + $39,906.00* = $122,532.00.

   Cross-flow: 4 cells with 60 horsepower drivers = 240 HP.
   Cost = $82,629.00 + $16,784.00 = $99,413.00.

2. **PUMP MOTORS**

   Counter-flow: 30 foot pumping head.

   $$BHP = \frac{(20{,}000)\ (30)}{(3960)\ (.80)\ (1)} = 190\ HP\ (150 + 40).$$

   $61,570,00 + $16,784.00 = $78,354.00.

   Cross-flow: 60 foot pumping head.

   $$BHP = \frac{(20{,}000)\ (60)}{(3960)\ (.80)\ (1)} = 380\ HP\ (200 + 150 + 30).$$

   $82,629,00 + $61,570.00 + $12,518.00 = $156.717.00.

3. **TOTAL YEARLY CONSUMPTION**

   Counter-flow: $200,886.00.

   Cross-flow: $256,130.00

The Counter-flow tower utilizes $55,244.00 less a year. On a ten year basis, it will save $552,410.00 in electricity at a constant 5¢ per KwH cost but with estimated yearly utility increases, it should be over $1,000,000.00 less than the Cross-flow tower in power consumption. This, of course, makes the Counter-flow tower the better selection even though it might have a higher first cost purchase price

*Motor HP and operating costs per KwH are listed on Fig. 7.

Counterflow and Crossflow Cooling Towers

. . . design concepts

ALL COOLING TOWERS except the small packaged models are custom made to meet a particular set of thermal parameters. Industrial cooling towers are available in two basic operational designs—counterflow, Fig. 1, and crossflow, Fig. 2.

Components—The basic components are the same in crossflow and counterflow towers; arrangement of parts may differ.

- Mechanical equipment—Typically, both types have a propeller fan connected to a direct-drive motor through a mechanical shaft and coupling; each has a gear-box drive train. Belts, pulleys, and centrifugal blowers may be used in some designs.
- Water-distribution system—Counterflow towers disperse water through a low-pressure spray-distribution system from a pipe running across the entire tower; in older models, a trough arrangement is used. Crossflow designs have open gravity-type hot-water basins above the fill.
- Fill—Baffling within the tower provides large water surface areas for heat transfer. Two classes of fill material are used: splash bars made from wood, metal, transite, or plastic; and film pack (cellular fill) constructed from thin sections of asbestos impregnated with chlorinated rubber, PVC, or polypropylene. The splash type uses entering air to cool the water droplets as they bounce down a series of slats; cellular-fill units convert the droplets into a thin film to provide a larger surface area for heat transfer.
- Drift eliminators—After the air has interacted with the falling water, baffling is again encountered to remove entrained moisture. The air usually changes direction about three times before it is exhausted from the stack. Counterflow-tower drift eliminators are installed horizontally between the water-distribution system and the fan; crossflow eliminators are placed vertically between the fill and center plenum.
- Cold-water collecting basin—As the hot water moves through the fill, it gives up heat to the passing air. Cooled water collects in the bottom basin, recirculates through the heat source, and returns to the tower to drop its heat load again.
- Air-intake louvers—Air is di-

Fig. 1. Counterflow towers have air-intake louvers at the bottom of one, two, three, or four sides, depending on design conditions. Air enters horizontally, immediately turns upward, and contacts water falling from a low-pressure spray-distribution system. Water moving through the tower strikes splash bars or cellular fill to accelerate the heat-exchange process. Moisture-laden air rises through the distribution piping, past the drift eliminators, and into the plenum chamber at the top. The fan exhausts the air through the stack.

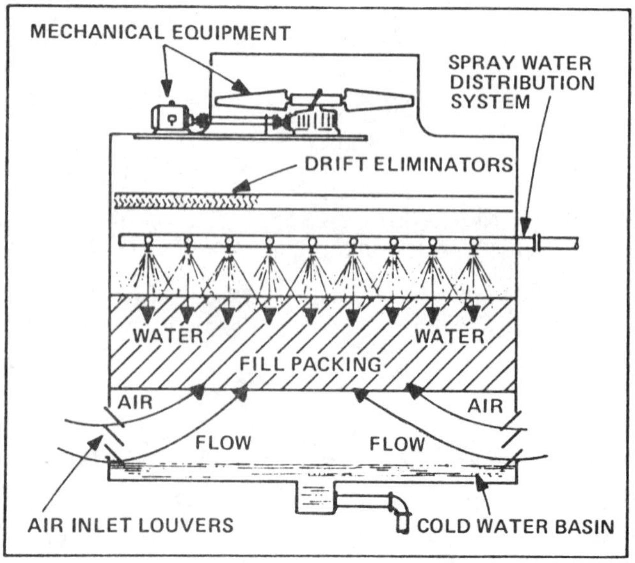

Fig. 2. Crossflow towers have air-intake louvers located along the entire height of the sides. Air moves horizontally through the fill while water falls vertically, by gravity, from an open distribution basin on top of the tower. Air moves through the drift eliminators into a plenum chamber in the center of the tower. The fan draws the air out of the tower.

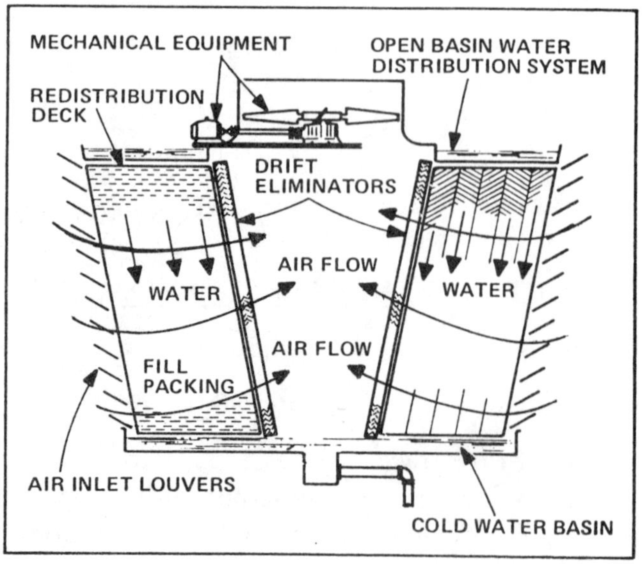

rected through the fill by louvers. Air enters at the bottom of a counterflow tower and through the entire face of a crossflow unit.

- Redistribution deck—Crossflow towers require additional decking to break up the solid columns of water dropping from the open basin into the fill area. In some instances, nozzles are added to improve the gravity distribution.

Air/Water Interaction—The relationship between the falling hot water and the incoming air determines the tower classification.

In counterflow towers, air-intake louvers are at the bottom. One, two, three, or four sides may be used depending on design conditions. Air enters horizontally and immediately turns upward to contact water falling from a low-pressure spray-distribution system. Water moving through the tower strikes splash bars or cellular fill to increase the heat-exchange rate. Moisture-laden air rises through the distribution piping, past the drift eliminators, and into the plenum chamber at the top of the tower. The fan exhausts the air to the atmosphere through the stack.

With this tower configuration, the cooler ambient air strikes the coldest water first. As the cold air rises, the water temperature decreases and heat exchange is accelerated. About 25 percent of the cooling is accomplished by this exchange of sensible heat; about 75 percent is produced by evaporation.

Packaged counterflow-tower capacities range from 5 to 1000 gpm; field-erected equipment capacities can exceed 20,000 gpm. Large towers are created by assembling cells of uniform cross section. The smaller units may have a centrifugal blower at the base to drive air up through the fill (forced draft), larger sizes have propeller fans at the top to draw air through (induced draft).

Crossflow towers have air-intake louvers along the entire height of the sides. Air moves horizontally through the fill while water falls vertically, by gravity, from an open distribution basin on top of the tower. Air moves through the drift eliminators into a plenum chamber in the center of the tower. The fan draws the air out of the tower.

In this design, air entering at the bottom encounters the coldest water, while that coming in near the top contacts the hottest water; therefore, water temperature will vary significantly by position within the tower.

Small packaged crossflow tower capacities start at 15 gpm, and air enters on only one side. Packages up to 2500 gpm normally have air entering on two sides. Crossflow tower capacities may also exceed 20,000 gpm. All towers of this type are induced draft; propeller blades up to 30 ft in diameter are used.

Figure 3 shows a crossflow (right) and a counterflow tower installed side by side.

Selection Considerations—With the same thermal conditions, size, and quantity of fill, a crossflow tower will require more air than a counterflow type. However, less static pressure is encountered in the crossflow because of its lower profile, and operating and mechanical-equipment costs are lower. The water flow rate of this type can be easily altered.

The crossflow design has a larger plenum area and is more susceptible to biological attack. Its open-pan distribution system allows the development of slime and algae, which must be controlled with chemicals.

The counterflow tower's closed-pipe distribution system is more difficult to clean. These units tend to be higher than crossflows, but not as long, so they need less land.

Crossflow towers have dominated the industrial market in recent years. But it is usually advisable to provide the manufacturer or consultant with the basic design requirements and let him make the calculations for both types to see which best meets the plant's needs.

Fig. 3. Some plants use both counterflow (left) and crossflow (right) cooling towers as heat-rejection devices.

CHAPTER 4

WATER DISTRIBUTION

Cooling Tower Water

...distribution and elimination

The function of a cooling tower, by obvious nomenclature, is to cool water — whether it is for compressors, air conditioning, chemical processes, power generation or any other related thermal problems.

The key to performance is *UNIFORMITY* for the three elements of any cooling tower's operation — water, air, and wet decking fill.

This article will discuss water distribution, together with its correlated effect, drift (water) elimination.

Upgrading performance of cooling towers economically utilizing existing mechanical equipment and water loading were covered in articles appearing in the October 1971 and November 1970 issues of the NATIONAL ENGINEER, reprints of which are available.

Counter flow (Fig. 1) forced draft towers (Fig. 2) only will be investigated since these structures lend themselves very well to the techniques described while Cross-flow towers (Fig. 3) with their gravity feed thru carefully calculated orifices pierced in the hot water distribution basins cannot be altered without changing the wet decking design and interior tower configuration. This was examined in depth in a previous article.

The types of water distribution systems vary with the number of design engineers producing blueprints, the most common being low pressure, small orifice (1/8″) metal nozzles (Fig. 4). Since the spray cone developed by this traditional shaped spout is rather small, a relatively large number of nozzles must be utilized to

FIG. 1 — (Left above) Counter flow tower for public utility recently upgraded proved to be a 12% improvement.

FIG. 2 — (above) Package counter flow forced draft tower lends itself to conversion.

FIG. 3 — (Left) Cross flow tower requiring special techniques for rebuilding (discussed in earlier NE article by author).

FIG. 4 — Standard small orifice metal nozzle tower in the process of rebuilding.

FIG. 5 — Gravity flume water distribution together with old style wood drift eliminator blades.

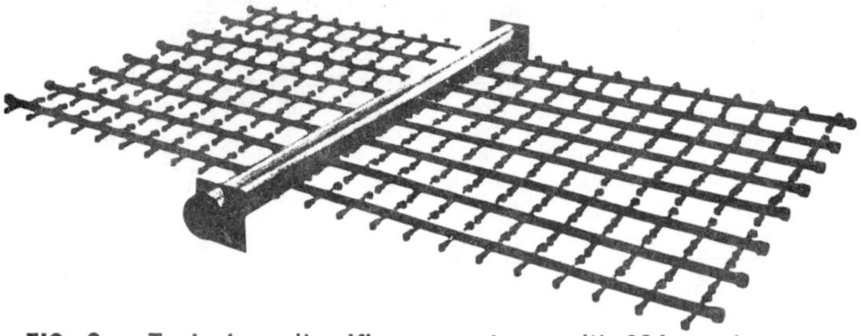

FIG. 6 — Typical small orifice spray trees with 384 nozzles on ¾″ iron pipe arms which are prone to clogging.

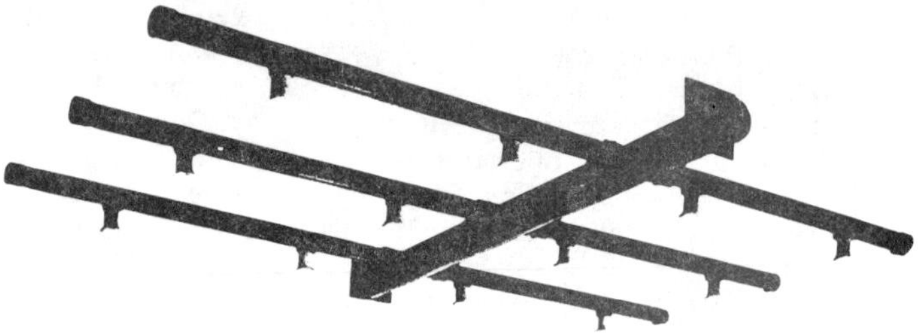

FIG. 7 — Non-corroding PVC piping, 3″ arms and 12 ceramic nozzles replace Fig. 6 but handling same GPM of water.

produce uniform water dispersement over the wet decking fill. There is a moderately straight line graph relationship enlarging the pipe header and distribution arm diameters with the nozzle orifice diameter to conduct higher volumes of water through the system. The necessary hydraulic calculations must be made to determine the precise relationship — but the obvious direction is apparent — the more GPM required, the bigger the pipes, nozzles, and tower required.

There are still in operation, many of the older type gravity flume distribution systems (Fig. 5). As the illustration shows — many things besides dirt, leaves, and debris get into this open system.

A salesman's stock in trade basically, is words, but a good sale must be a two-way street, i.e., what is in it for both sides?

If a given cooling tower is performing well, if the pipes are not corroding, if the nozzles are not clogged up and if the maintenance costs are negligible, then there is no requirement to spend money on the installation. On the other hand — if better performance thru colder water is urgent, if the maintenance department has to spend many man hours cleaning nonfunctioning nozzles (Fig. 6) and management is considering spending capital investment money for new facilities, now is the time to investigate the possibility of upgrading the existing plant by changing the water distribution to a highly efficient method — utilizing cheaper tax deductible maintenance dollars.

Experience and many installations have proven the high efficiency of water distribution of the specially designed ceramic nozzle, together with non-corroding Polyvinyl Chloride (PVC) piping (Fig. 7). Contrasted to Fig. 6 which was designed for the same GPM of water, it is obvious that the labor savings in not having to clean the 12 ceramic 1″ orifice units periodically will more than offset their installation cost, against the old fashioned ⅛″ openings on the 384 nozzles. Figs. 8 and 9 illustrate this fact on an actual field change. Also, the large diameter 3″ PVC piping will not clog with rust readily, and if by chance it does — it is simple to unscrew the end cap for blow down. The nozzle, incidentally (Fig. 10) snaps out of a baynet slot holder by hand pressure.

The above portion has been devoted to small package units; however, the same techniques apply to larger field erected installations (Fig. 11).

The necessary hydraulic calculations will determine the pipe sizes and orifice openings. Fig. 12 shows a 14″ diameter pipe assembly with 6″ dia. arms while Fig. 13 illustrates the nozzles on six foot centers vs. 36″ for Fig. 9. The water distribution pattern on Fig. 14 indicates a slight overlap of the cone in the previous installation. This depth is calculated and located at the optimum point by the manufacturer's curves and charts.

If existing steel header pipes and arms are corroded to the point of replacement, the labor cost of installing PVC piping over galvanized steel is less. A man with a rope can hoist the plastic pipe and have an assistant help lock it in place. The weight of steel will require riggers and a large team of men to do the same thing.

Water volumes from 15 GPM to 250 GPM per nozzle are available in the 2 to 7 psi pressure range which are well within the operating parameters of today's cooling tower requirements.

DRIFT ELIMINATION is a problem caused by the nozzles turning the water into fine droplets, some of which are carried upwards by the vertical air stream, before the water becomes impinged upon and into the fill.

Standard sections fabricated of as-

FIG. 8 — Field change of corroded and non-functioning small nozzles.

FIG. 9 — Spray system of PVC and 12 ceramic nozzles replacing Fig. 8.

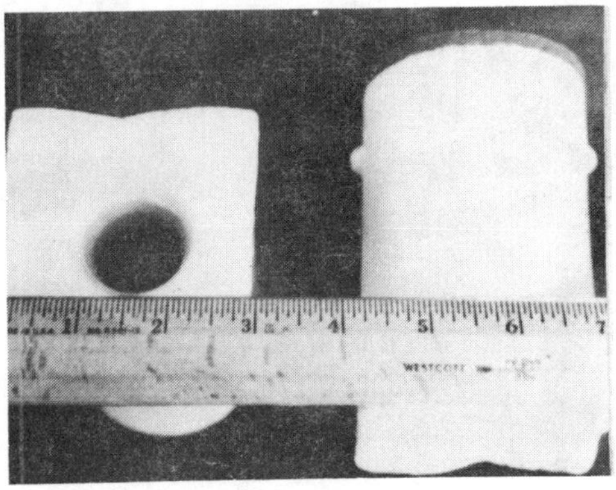

FIG. 10 — Ceramic non-clogging non-corroding nozzle.

COOLING:

bestos cellular honeycomb baffles in aluminum or stainless steel framing are available (Fig. 15) to take the place of corroded steel units in package towers Figs. 2 and 4. The larger field erected towers Figs. 16 and 1 usually have heavy wood slats, these are partially illustrated by Fig. 5. This

FIG. 11 — Plastic pipe 12″ diameter header easily installed by minimum manpower.

FIG. 12 — Stainless bolts on 14″ pipe flanges and 6″ arms.

FIG. 15 — Standard aluminum or stainless steel encased fireproof, efficient drift eliminator.

FIG. 16 — Typical power generating plant counter flow tower of 18,000 GPM capacity is easily upgraded.

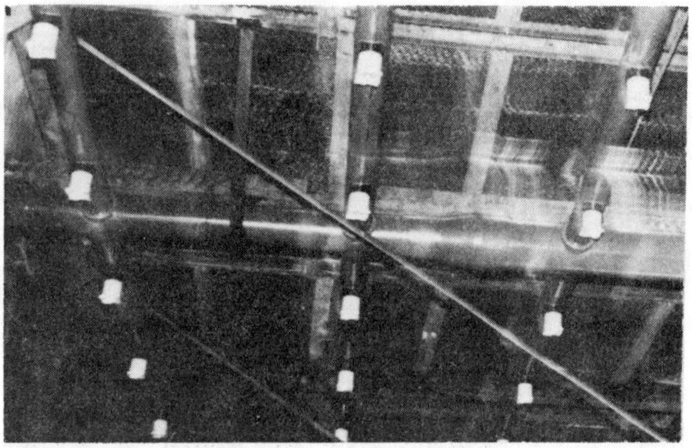

FIG. 13 — Nozzles on 4′center are adequate for 12,000 GPM distribution.

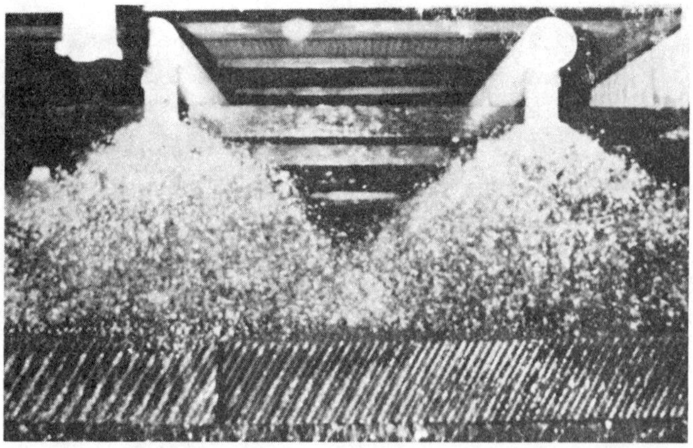

FIG. 14 — Spray cone for 4000-ton capacity tower. Nozzle is adjustable for 40-ton water.

FIG. 17 — Large drift eliminator installation can be walked on without harm to surface.

FIG. 18 — Chemically resistant conversion of PVC piping, drift eliminator and wet decking. Spray nozzles are ceramic; steel is epoxy-coated.

FIG. 19 — (BEFORE) — Owner considered junking tower and installing new unit.

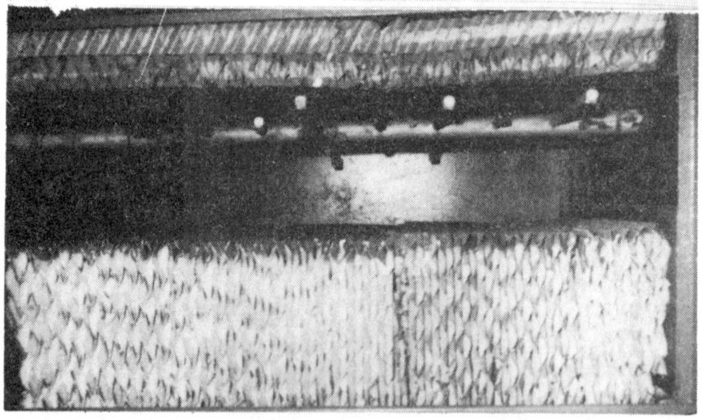

FIG. 20 — (AFTER) — Prior to installation of PVC piping at $8,-000 savings for 100-ton tower.

old-fashioned baffle is inefficient and after some walking on top of the wood by maintenance mechanics, permits excessive water escape thru the broken wood.

The logical replacement for the heavy wood sections is shown in Fig. 17. The expanded aluminum or stainless steel mesh is more than sturdy enough to support a work load, and a standard 4″ thick section will contain the accepted restraint, permitting less than ½ of 1% of water thru up to an air velocity of 750 FPM.

Most counter flow towers work in the area of 350 to 450 FPM. At a slight fractional addition of static pressure, a 6″ thick eliminator will perform at velocities up to 975 FPM.

An added bonus of using this type of installation is that the fireproof asbestos construction acts as a fire stop to prevent burning cigarettes, rags or flame from setting the crisp redwood on fire.

For special chemical plant process installations, a tower composed of PVC piping, ceramic nozzles, PVC drift eliminators and PVC wet decking Fig. 18 is available. Polypropylene, ABSPVC and other types of materials are also available. The cooling tower structure, of course, must be lined with a resistant rubber or epoxy coating.

The value of this type of conversion is illustrated in Figs. 19 and 20. In the before picture, the owner was going to scrap the tower and install a new unit. The after picture prior to new spray installation of only two nozzles, saved the owner in the area of $8,000.00 on this 100 ton unit. *And this is what the foregoing is all about!!*

Crossflow Redistribution

■ Basic principle of cross flow cooling tower operation is that while water drops vertically through wet decking, air travels horizontally, or crosswise, at 90° to water flow. By contrast, water drops vertically while air rises vertically, 180° counter to water direction, in counter flow tower. Counter flow water distribution is comprised of spray piping and nozzle system, discussed in the March 1976 Cooling Tower Technology article.

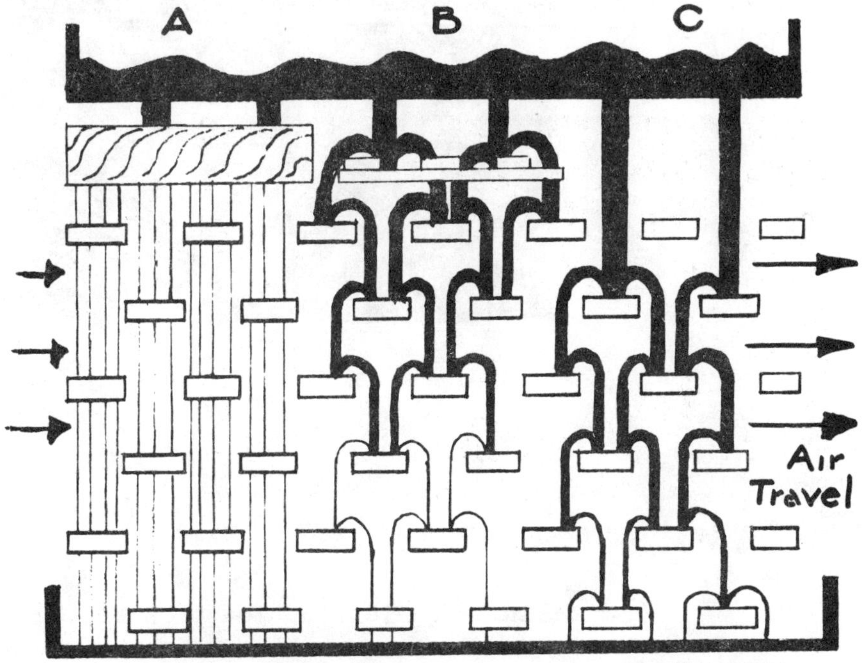

BROKEN, MALFUNCTIONING decks permit solid columns of water to drop up to 2′, or almost one-third the distance through the fill before randomly splashing into small droplets for cooling (as shown at right above)

Cross flow water distribution is by gravity from an upper hot water basin having a series of holes through which a calculated volume flows. Object is to break solid column of water into droplets as soon as possible so that flowing air can cool these droplets as they bounce from splash bar to splash bar before reaching the cold water basin.

Until recently, a lattice work of thin wood slats was placed beneath the solid, flowing column of water (Fig. B). As water splashed at random on this redistribution deck, it was gradually broken into smaller columns and finally into droplets. After a few years of use, maintenance, and abuse, these decks are broken and sometimes removed since their true function is not well understood. The result, in some cases, is water dropping a third of the way through the fill before being broken into small enough drops for proper and effective cooling, (see schematic).

Cellular redistribution packs are now being utilized to break the water columns into droplets within 4″ of the hot water basin, obtaining maximum cooling effect of the air/water interface (Fig. A).

It is relatively simple to temporarily take out the air stops of the back fan side of the fill and remove what remains of the old-fashioned, broken wood redistribution decks. Custom-made cellular packs can then be put into position and water stops replaced to prevent extraneous spray from being drawn up by the fan.

Installation of these redistribution decks, fabricated of 100% fireproof materials, can in some cases improve performance of an existing cooling tower as much as 15% at small cost.

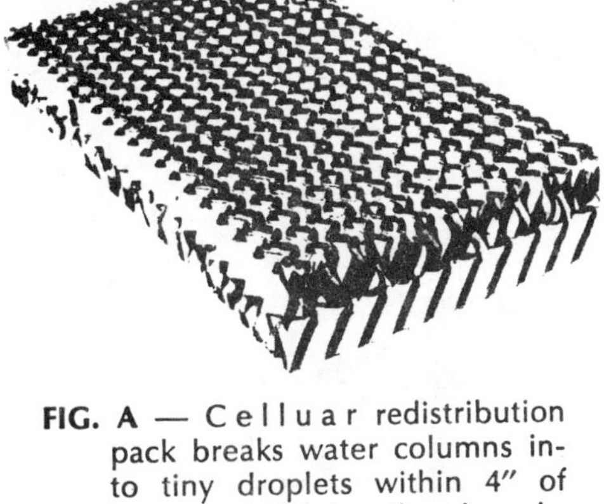

FIG. A — Celluar redistribution pack breaks water columns into tiny droplets within 4″ of hot water basin, thereby obtaining maximum cooling effect of air/water interface (as shown in left portion of schematic above).

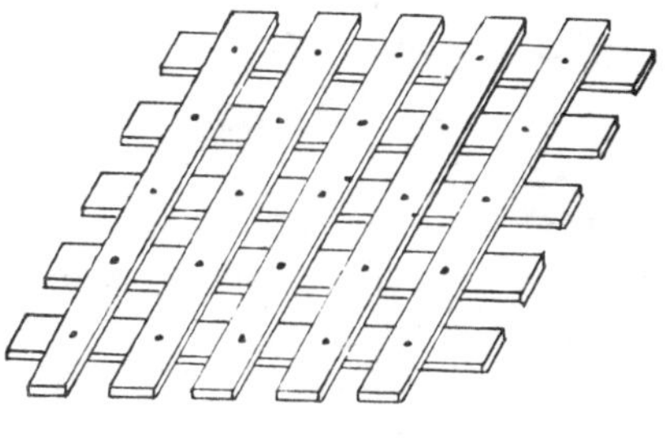

FIG. B — Standard well-repaired redistribution lattice desks breaks water into drops within 14″ to 16″ for cooling (also shown in B, center, portion of schematic above).

Cellular redistribution pack breaks water columns into tiny droplets within 4" of hot basins thereby obtaining maximum utilization of cooling effect of the air/water interface.

CHAPTER 5

DRIFT ELIMINATION

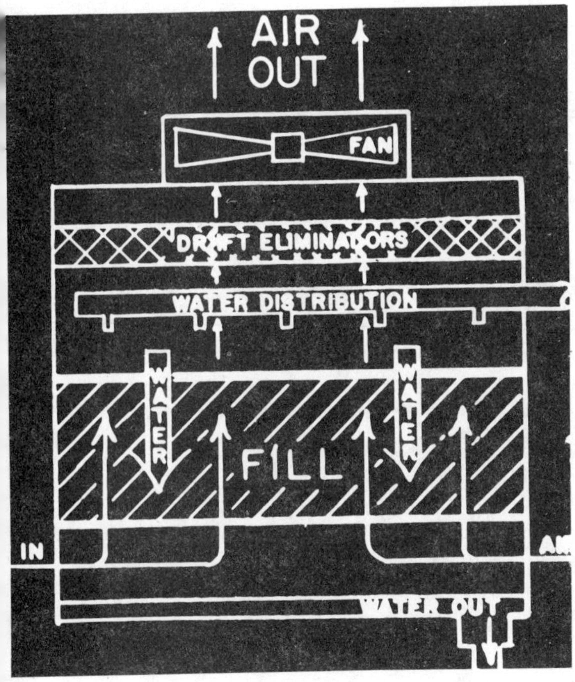

Figure 1. Schematic of counter-flow design.

Figure 2. Schematic of cross-flow design.

Cooling Tower Drift Elimination

Tests with cooling towers indicate that conversion to cellular wet decking fill and drift eliminators can, without changing horsepower, improve operating efficiency by more than 30%.

A cellular drift eliminator has been developed for cooling towers that will prevent water loss, prolong tower life, and provide improved plant efficiency. Before discussing drift elimination in detail, however, we will take a brief look at the principles of cooling towers and at some of the equipment designs that employ them.

Mechanical draft counter-flow and cross-flow cooling towers are used for industrial process water, air conditioning, refrigeration, and electric generating plant condenser cooling. The two basic tower configurations are so identified by the character of the air-water mix. In the older, counter-flow design, shown schematically in Figure 1, the air and water mix counter to each other, vertically at 180° differential. Water is sprayed or distributed downward through the tower fill, and the air is driven or induced upward directly against the descending water. Vertical air velocity through this cooling tower can vary from 300 to 700 ft./min.

The cross-flow tower, shown schematically in Figure 2, causes the air and water flows to cross each other at 90°, with the air travelling horizontally through the vertically descending water. In this case, air is induced through the wet decking fill horizontally and then changes its direction at discharge. Here again, a baffle system is necessary as the air, which in cross-flow towers range from 150 to 750 ft./min. velocity, passes through the flooded zone and out the fan cylinders.

A baffle, or drift eliminator system, is placed between the water distribution system and the air discharge point to minimize dispersal of entrained water droplets into the surrounding atmosphere.

Four factors in efficient operation

Drift eliminator efficiency is an essential element of cooling tower design for many reasons, among them: 1) conservation of water; 2) retention of chemical treatment in the cooling water; 3) prevention of staining by chemical additives, as chromates; and 4) avoidance of violation of local area environmental protection regulations.

Cooling tower performance, i.e., heat removal, is a balance between water flow and air volume; therefore drift eliminators are normally designed to be efficient through a calculated range of air flow. Too great an air speed can result in excessive drifting of water from the tower, while poorly designed eliminators will increase static pressure through the tower, slowing the air and lowering the cooling capabilities and efficiency of the unit.

The need for an efficient mechanical system to retain the water is obvious since the spray or water distribution system turns the heavy stream of condenser or process water into light droplets preparatory to being cooled by the airstream. The evaporative cooling is generated by the mixture of air and water, plus loss of sensible heat. As the air moves counter or cross flow to the water, it will pick up much of the mist and droplets, and carry them with the air flow out of the tower.

The schematic operation of a three-pass drift eliminator is shown in Figure 3. A newer, more efficient drift eliminator fabricated from asbestos and impregnated with chlorinated rubber has been developed. It causes the air to gently change its direction approximately six times, thereby obtaining greater surface contact to release

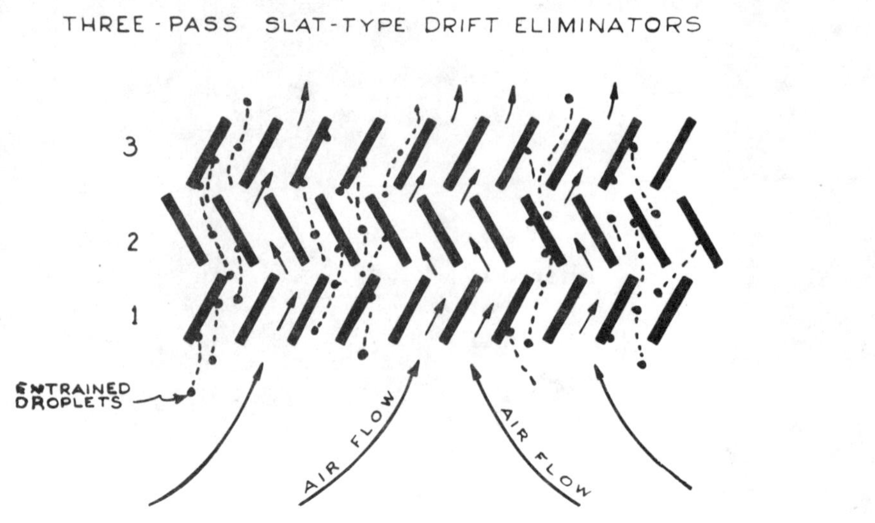

Figure 3. Schematic of wood blade drift eliminator.

Figure 4. Schematic of cellular fill drift eliminator.

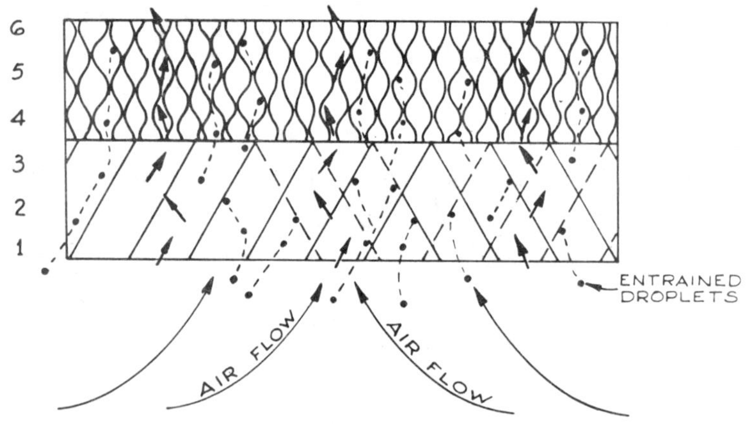

water droplets. Figure 4 illustrates the schematic mechanics of the six-pass cooling.

The curves in Figure 5 are pressure drop (in H_2O) plotted against air velocity. They were obtained from two manufacturers' elimination systems comparing pressure drop of conventional three-pass wood sections with the baffling of cellular sections. This graph indicates there is less power consumption for more work being done. Calculating actual monetary gains is difficult, but it is obvious that a freer flow of air will provide considerably more cooling at less cost on a straight line relationship. The pressure drop curves show that the new efficient cellular drift eliminators are about 7% more efficient than three-pass wood.

If, for example, one cell of a cooling tower required 50 h.p. to produce an 18° cooling range, and a 10° approach for 1,000 gal./min. of water, this year in the northeast, a utility company will charge approximately \$11,450 to keep this motor operating continuously. Therefore, if this cell can be rebuilt to cool 1,500 gal./min. to the design criteria, a 50% improvement can be obtained. The \$11,450 power charge then becomes \$5,725 to cool the original 1,000 gal./min.

Older towers can, and have been, upgraded. Installation of cellular fill, polyvinyl chloride piping with ceramic nozzles, and cellular drift eliminators have achieved a 22% improvement on original design condi-

Figure 5. Pressure drop curves illustrating higher resistance of old fashioned wood blades vs. new cellular eliminators.

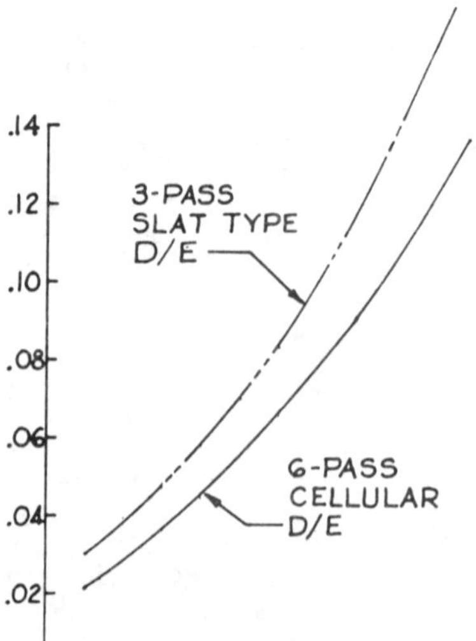

Figure 6. Deteriorated wood drift eliminators of a power plant cooling tower.

tions (actually never obtained in the first place), and 40% efficiency over the previous year's operation.

Using this example, it follows that a 7% efficiency improvement of cellular fill translates itself into $801.50 saving, or the ability of enabling the tower to cool a 105 gal./min. share of the additional 1,500 gal./min. load in the above example.

One key to the rebuilding and upgrading procedure is that the considerably lower static pressure resistance of the cellular drift eliminators permit the available air to flow at a higher velocity at no additional cost. The versatility of cellular drift eliminator design can be seen in its applications to varying conditions.

Figure 6 shows the deteriorated wood drift eliminators of a power plant cooling tower. This complex was not producing design cold water, but it was producing excessive drift which led to the formation of puddles on the deck of the facility. Figure 7 shows the same installation after cellular drift eliminators had been installed. The "before" photograph, Figure 6, shows water droplets being drawn out of the tower, while in the "after" photograph it can be seen that, even though the fan is rotating, all of the water droplets are constrained and only a slight fog is leaving the fan cylinder.

A common method of trying to improve cooling tower performance is to increase the air velocity by rotating the fan with more power and higher pitch angle to the blades. This often causes increased drift loss because the eliminators were not originally designed for this type of service. Cellular drift eliminators which do not increase the static pressure, but will increase the drift collecting capability, can minimize such an eventuality.

In fact, reduced static pressure resistance permits more air through the tower at the same horsepower expenditure, but it will permit more work at the same cost before the conversion.

Complete drift elimination is possible in theory, but impractical in application. Acceptable level generally satisfactory is "not in excess of 0.2%."

Cross-flow towers are often susceptible to drift problems, because of a tendency toward higher air flow velocities in this type of design. Since the drift requirements are more stringent, a different type of eliminator system is used with a very delicate balance design and performance. If one or two eliminator blades are out of position, drift will occur quite readily. Several manufacturers have developed cellular eliminators to overcome the problem. Here again, the lower resistance of these new types of eliminator, plus their drift elimination effectiveness, make them a desirable conversion from the less efficient wood.

Figure 7. Cellular drift eliminators retain mist.

Economic data

A cost analysis is difficult, because the efficiency, or lack thereof, of water cooling equipment affects many phases of the operation. If the cooling tower will reduce by even one-half of one degree the water temperature at the lower end of the approach to the wet bulb, the compressors and condensors can work more efficiently and produce better results at lower cost.

Actual field experience and testing of cooling towers indicates that a conversion to cellular wet decking fill

Figure 8. Cellular eliminators with PVC spray underneath.

Figure 9. Chlorinated rubber impregnation prevents atmospheric attack on cellular eliminators.

and drift eliminators, without changing the motor horsepower, can improve cold water temperatures as much as 3°F. This increase in a 10°F. range provides over a 30% improvement. Drift eliminators alone, depending upon efficiency and pressure drop through the existing old installation, cannot be expected to provide more than one-quarter to one-half degree improvement. To translate one-half degree into dollars and cents, one user found that the $25,000/yr. cost to power his cooling tower was reduced by 5%, a $1,250 saving. This is not the only significant factor however. Savings in make-up water, make-up chemicals, and compliance with local ordinances will add up to more economical operations for the owner.

While a water trough distribution system is antiquated and extremely inefficient, there are many installations still using this flume configuration. It is quite obvious that when the water is agitated, droplets can be drawn up by the fan together with water induced through the poor elimination system. It can be seen in Figure 8 how this is solved by the installation of cellular eliminators above the distributors, which in this case are bound with stainless steel and protected with stainless steel expanded mesh.

Many cross-flow towers were originally constructed with a single or double-pass herringbone baffle at the fan side of the unit. A cellular drift elimination installation adds to the efficiency of the operation by stopping the air from making heavy turns around the herringbone, and also collects any droplets which may be pulled towards the center by the fan.

Many packaged cooling towers used a zig-zag galvanized steel assembly for drift elimination. Besides being restrictive to air flow, excessive drift develops as the water works its way up along the metal surface of the eliminator blades. Sulfuric fumes, prevalent in most city atmospheres, are drawn inside the cooling tower and convert to sulfuric acid, which corrodes galvanizing and results in blades which deteriorate.

The deterioration process is quite evident in the picture of a roof top cooling tower, Figure 9. It is obvious that replacement is necessary, and after the owners opted to install cellular eliminators which are guaranteed against deterioration for five years, this problem will not recur despite a harsh sulfuric atmosphere.

An additional bonus of neoprene-asbestos material is that it provides a fireproof barrier. Thus, a match, lighted cigarette, or burning rag from an incinerator that may land on top of the drift eliminators when the tower is not in operation, will not start a conflagration because the fire will extinguish itself on the asbestos and not reach the flammable wood below.

CHAPTER 6

THERMAL EVALUATION

THERMAL EVALUATION

Many theories have been developed describing the heat and mass transfer which takes place inside cooling towers. Most of these theories are based on sound engineering principles since the cooling tower may be considered as a heat exchanger in which water and air are in direct contact with each other.

There is no acceptable method for accurately calculating the total contact surfaces between the water and air inside the tower. Therefore, a "K" factor or heat transfer co-efficient cannot be determined directly or calculated by known heat transfer theories.

Many types of heat mass transfer devices defy solution by theoretical methods. Design data must be obtained by full scale tests under actual operating conditions. The results can then be plotted and formulated.

"Black Box" Theory of Experimentation

Experimentation and research utilized in analyzing cooling towers may be described as the "Black Box Theory". That is, the boundary conditions are defined and described by fundamental equations and mathematical modeling. The exact process by which the heat and mass transfer takes place within a direct contact heat exchanger is not fully known. Figure 1. However, the results may be formulated into equations which adequately evaluates and predicts performance of the device under most operating conditions.

The Merkel Theory

The generally accepted concept of evaluating water cooling performance was developed by Merkel in 1925. His analysis and equations included the sensible and latent heat transfer into an overall heat and mass transfer process based on enthalpy difference as the basic driving force.

FIGURE 1
"BLACK-BOX" THEORY OF EXPERIMENTATION

WHERE:
- L = Liquid (Water)
- G = Gas (Air)
- T = Temperatures
- M_L = Mass of Liquid (GPM)
- M_G = Mass of Gas (CFM of Air)

Each water surface is surrounded by a film of saturated air at the bulk water temperature. Figure 2. The air becomes heated and saturated as it passes through the cooling tower. The temperature and associated physical properties of both water and air change according to their relative position.

The heat and mass transfer have been related to the enthalpy difference at each point or each temperature increment. The Merkel Theory, Figure 2, established that atmospheric water cooling is basically related to and limited by the wet bulb temperature which is an approximate measure of enthalpy. The wet bulb is determined from the psychrometric chart and defined as the intersection of the ambient temperature and dew point or ambient temperature and relative humidity. The wet bulb is an indication of the availability of evaporative potential of the atmosphere. The wet bulb is also referred to as the temperature of adiabatic saturation. Water cannot be cooled to below the wet bulb temperature in a cooling tower.

The KaV/L equation is a determination of the counterflow thermal demand while the KaY/L equation is a determination of the crossflow thermal demand. This is called the "tower characteristic" which is independent of the type and configuration of fill packing. The values required to obtain the KaV/L or KaY/L are based on the required design conditions of the particular heat rejection situation being investigated.

FIGURE 2.

MERKEL THEORY

COUNTER FLOW TOWER PERFORMANCE

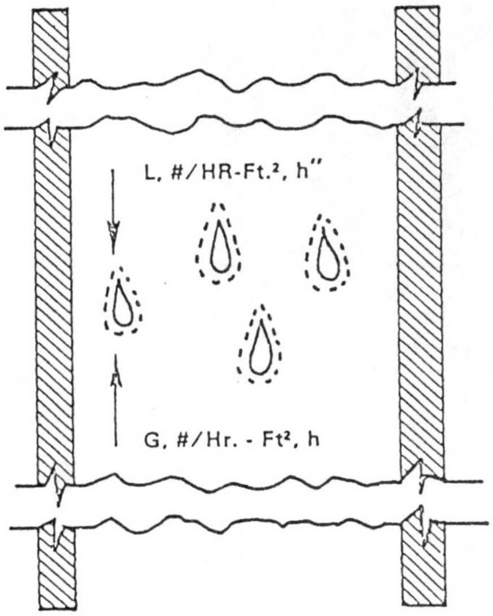

LICHTEN/STEIN 1943 CALCULATIONS

1. $$\frac{KaV}{L} = \int_{T2}^{T1} \frac{dt}{h''-h}$$

2. DRIVING FORCE IS RELATED TO h''–h, OR THE ENTHALPY DIFFERENCE BETWEEN AIR SATURATED AT THE WATER TEMPERATURE (FILM SURROUNDING THE WATER SURFACE) AND THE ENTHALPY OF THE MAIN AIR STREAM.

3. TRUE DRIVING FORCE IS TEMPERATURE DIFFERENCE BETWEEN WATER AND AIR.

Kelly's Handbook of Crossflow Performance Curves, Figure 3

In 1976, Neil W. Kelly published Performance Curves for Crossflow Cooling Towers applying the basic formulas of the Four Point Tchebycheff Integration Method, but modified for the particular requirements of this type of crossflow tower configuration. His book, with its introduction to the performance curves, have dispelled much of the "mystery" surrounding crossflow technology and provides a sound guide for designers.

Figure 4 is a representation of a Kelly crossflow cooling tower performance curve of a constant 82°F inlet wet bulb and a 30°F cooling range. The Kelly curves extend from 35°F to 85°F covering many cooling ranges. The curves are plotted with the Thermal Demand KaY/L (Ordinate) as a function of the Liquid to Gas Ratio (Abscissa) L/G. The Approach lines (cold water temperature minus wet bulb temperature) are shown as the parameters. The Demand Curves are determined independently of tower design and performance of any particular fill.

Figure 3

82 °F, WET BULB
30 °F, RANGE

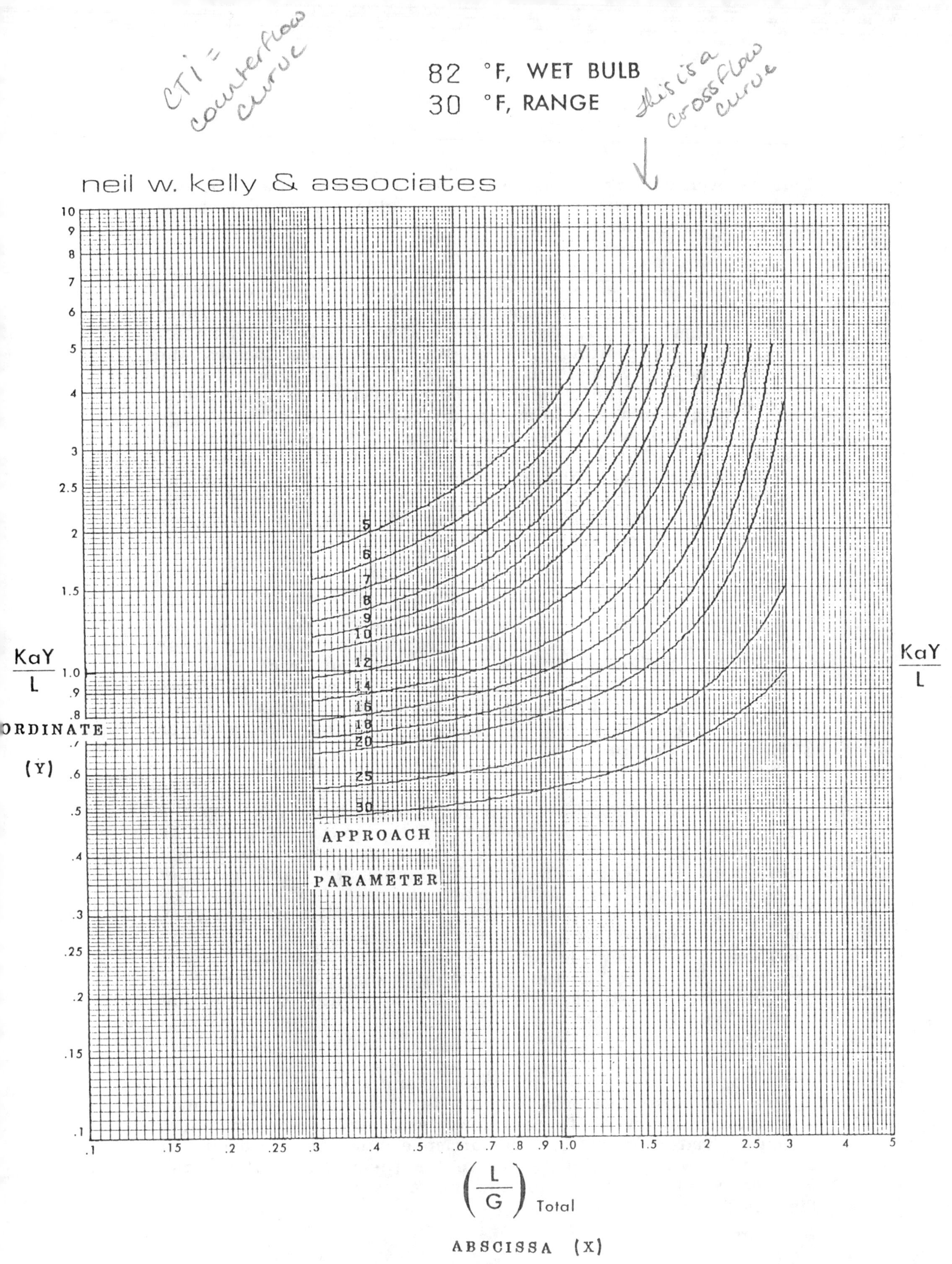

Figure 4

This reference book with the families of curves have become the accepted industry standard and should be on the desks of all cooling tower designers.

The cooling tower characteristic, KaY/L (crossflow) or KaV/L (counterflow) value is determined by the intersection of three points. Namely, L/G, Approach, and/or the Characteristic. Calculating and plotting any of the two values locates the third. While the characteristic can be laboriously computed by the application of the Tchebycheff intergration using enthalpy chart values, the accepted method of easily locating the Design Point is by calculating the liquid to gas ratio and intercepting the Approach Curve on the proper performance curve of design or test wet bulb and range. By thus plotting the Design Point at the intersection of L/G and Approach, this automatically locates the thermal demand KaY/L or KaV/L on the vertical ordinate scale.

Both the Tower Characteristic and Liquid to Gas ratio are dimensionless values. The larger the thermal demand on the ordinate scale, the more difficult it will be for a cooling tower to meet the design conditions. It is also apparent that the higher the air volume (the "G" value of the liquid to gas ratio) the lower the L/G ratio becomes as it moves to the left on the horizontal abscissa, the easier it will be for the tower to operate to Design.

The Thermal Demand, KaV/L or KaY/L must be equaled by the thermal performance and capability of the tower design. The intersection of the tower characteristic curve at the Design Point, with the Approach line determines the L/G ratio at which the tower must be operated to meet the specified or design temperatures. The design values of KaV/L or KaY/L and L/G, adjusted for changes in temperature may be transposed to other demand curves. The adjusted points predict the Approach for operating conditions other than design. This is an important relationship utilized in testing towers thermally.

Astute and knowledgeable purchasers and operators of cooling towers request manufacturers to provide performance characteristic curves of their tower's Design Point and heat transfer materials (fill) performance to be utilized in conjunction with the Kelly's Curve for their crossflow towers, for accurate predictions of the performance as depicted on Figures 5 and 8.

LIQUID TO GAS RATIO (L/G)

The L/G ratio on the Kelly Performance Curves are stated in logarithmic relationships and are notated as dimensionless values.

The Liquid, or waterflow (L) is specified in gallons per minute and the Gas (G), air volume is designated in cubic feet per minute. The mathematical problem is to find a common relation for both so that they will cancel out when one is divided by the other leaving a dimensionless remainder.

Recalling basic algebra, if you multiply the top of an equation and the bottom of the equation by the same quantity, you do not change the value of the equation. For example, if the top portion of any equation is multiplied by 3.14 and the bottom of that same equation is multiplied by 3.14, the result will be a higher value for both dividend and divisor, but it will not change the value of the relationship. Therefore, as we look at the example given below, if we multiply the top of the equation by 60 minutes and the bottom by one hour, since there are 60 minutes in one hour, the value of 1 does not change the equation. We know there are 0.1337 cubic feet per gallon of water and 62.06 pounds of water in a cubic foot, we can multiply the gallons per minute, with the L as a numerical value given in the design specifications. The various values cancel out giving the value of the water as L times 497.84 pounds per hour. The next step is to change the cubic feet per minute of air into pounds per hour which equals 4.262 pounds per hour so that as the air in pounds per hour is divided into the pounds of water per hour, they cancel out leaving a constant 116.8.

This is a common engineering exercise since, for example, the electrical constant for calculating brake horsepower vs pumping head is 3960 while the gas constant for BTU per pound - mole is 1.937.

LIQUID TO GAS CONSTANT CALCULATION

The constant for the liquid to gas ratio as explained above, is determined as follows:

Where L = numerical value of water in GPM.

G = numerical value of air in CFM.

$$GPM = \frac{L \text{ x gal}}{\text{min}} \times \frac{60 \text{ min}}{\text{hour}} \times \frac{0.1337 \text{ CUFT}}{\text{GAL}} \times \frac{62.06 \text{ lbs}}{\text{CUFT}}$$

$$= L \times \frac{497.84 \text{ lbs}}{\text{hour}}$$

$$CFM = \frac{G \text{ x CUFT}}{\text{min}} \times \frac{60 \text{ min}}{\text{hour}} \div \frac{14.078 \text{ lbs}}{\text{CUFT}}$$

$$= G \times \frac{4.262 \text{ lbs}}{\text{hour}}$$

$$L/G = \frac{497.84 \text{ lbs water/hour}}{4.26 \text{ lbs air/hour}}$$

Constant = 116.8

In actuality, the constant varies from approximately 115.9 to 117.1 depending upon altitude and temperatures. However, the average 116.5 is selected and the variation error can be easily absorbed by the width of a pencil mark on the performance curves since the curves are not that accurate to the third decimal place.

$\therefore$ Accepted average Constant = 116.5

Example:

Assuming an eight cell tower, with 104,000 GPM of circulating water and eight 28 foot fans driven by 200 horsepower motors, generating 1,476,000 CFM of air per fan, we calculate the L/G ratio for one cell, assuming that all cells are typical and perform in the same manner:-

$$L/G = \frac{\frac{104,000}{8} \times 116.5}{1,476,000}$$

$$= 1.03$$

Since this book has been written for cooling tower users of all types and sizes, while 104,000 GPM is a considerable amount of water, it is merely one cell of 13,000 GPM eight times. Or dividing by a factor of ten, the same procedure applies for a 1,300 GPM cell as well. Techniques and calculations are similar, regardless of the size.

Assume a one cell tower with 1,300 GPM of circulating water and one 12' diameter fan driven by a 40 HP motor, generating 125,000 CFM per fan, we calculate the L/G ratio in the above manner:-

$$L/G = \frac{1,300 \times 116.5}{125,000}$$

$$= 1.21$$

The next few pages will contain examples of fill selection in accordance with procedures that have been discussed.

FILL DEPTH CALCULATIONS WHEN DESIGN CONDITIONS OF THE COOLING TOWER ARE KNOWN AND AVAILABLE

Example:

I. Crossflow, cellular, less than 20 feet fill height.

Design Conditions

GPM:	22,000	
HWT:	95°F	12°F Range (HWT–CWT)
CWT:	83°F	11°F Approach (CWT–WBT)
WBT:	71°F	
CFM:	371,100	Per Fan

5 Cells
18 foot diameter fans
50 HP Motors

$$L/G = \frac{\frac{22,000}{5}}{371,100} \times 116.5 = 1.38$$

Step One

The initial inquiry will be to locate the Design Point on the particular performance curve for these Design Conditions. Figure 5. Opening up the Kelly Handbook of crossflow cooling tower performance curves, we will go to the 71°F wet bulb section and continue until we reach the 12°F range page. We know that the Approach to the wet bulb (CWT of 82°F minus WBT of 71°F) is 11°F. With a french curve, darken the 11°F Approach parameter half way between 10°F and 12°F to give us a visual base line.

Step Two

Now calculate the liquid to gas ratio as described on Page 5 which will work out to be 1.38. Traveling vertically on the 1.38 line, we then intercept the 11°F Approach at the Design Point. Figure 5. The design characteristic curve is a performance value that is furnished by fill manufacturers and is in the area of -.6 negative slope, or in plain language, approximately a 60° angle. This is drawn in and is used for our testing purposes. Locating the Design Point, we can then travel horizontally to the left hand KaV/L value of 1.52. Bearing in mind that the Design Point is the intersection of these three items.

Step Three

Select the fill manufacturer's performance curve that is likely to produce the results required. Figure 6. Since the cooling tower in this exercise has less than a 20' vertical height, cellular fill can still be utilized. All reputable manufacturers will furnish performance curves, which can be xeroxed into transparencies. Usually, in larger towers with a vertical fill height of more than 20', it is not economically viable to install cellular fill.

71 °F, WET BULB
12 °F, RANGE (82-71)

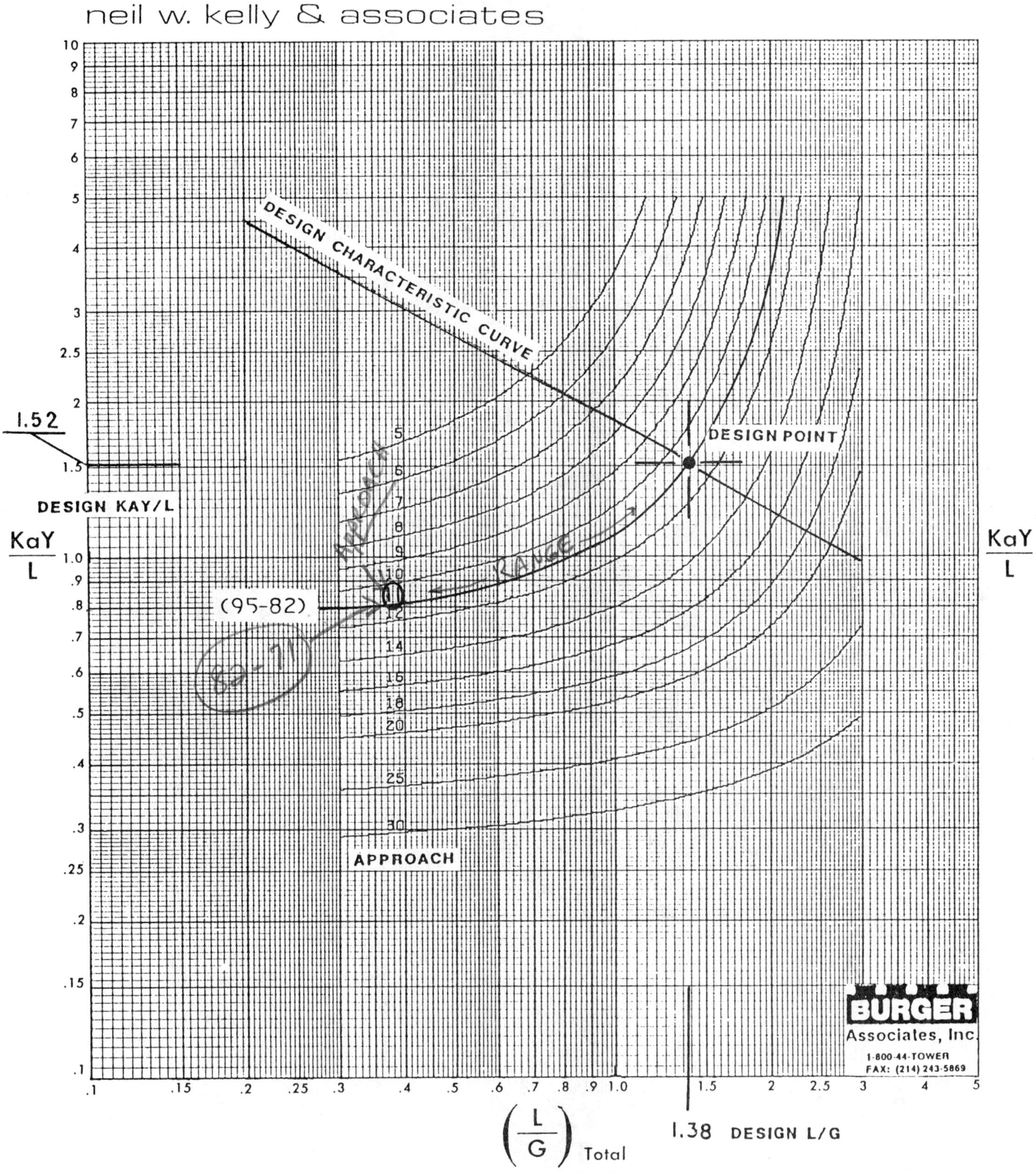

Figure 5

XF 19060 or 1900

Curves valid for all air velocities between 300 and 750 FPM

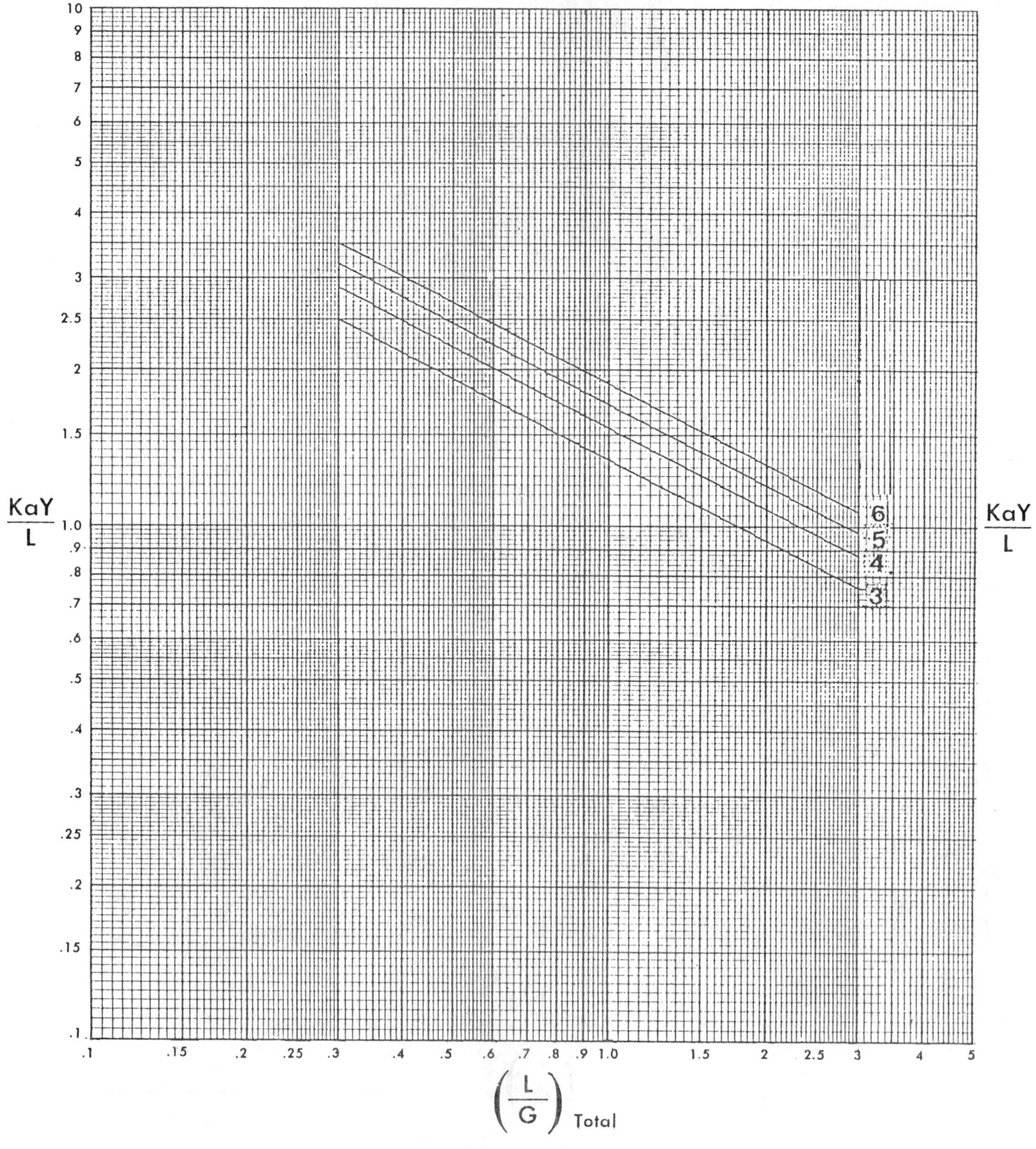

Figure 6

11,8,080,00

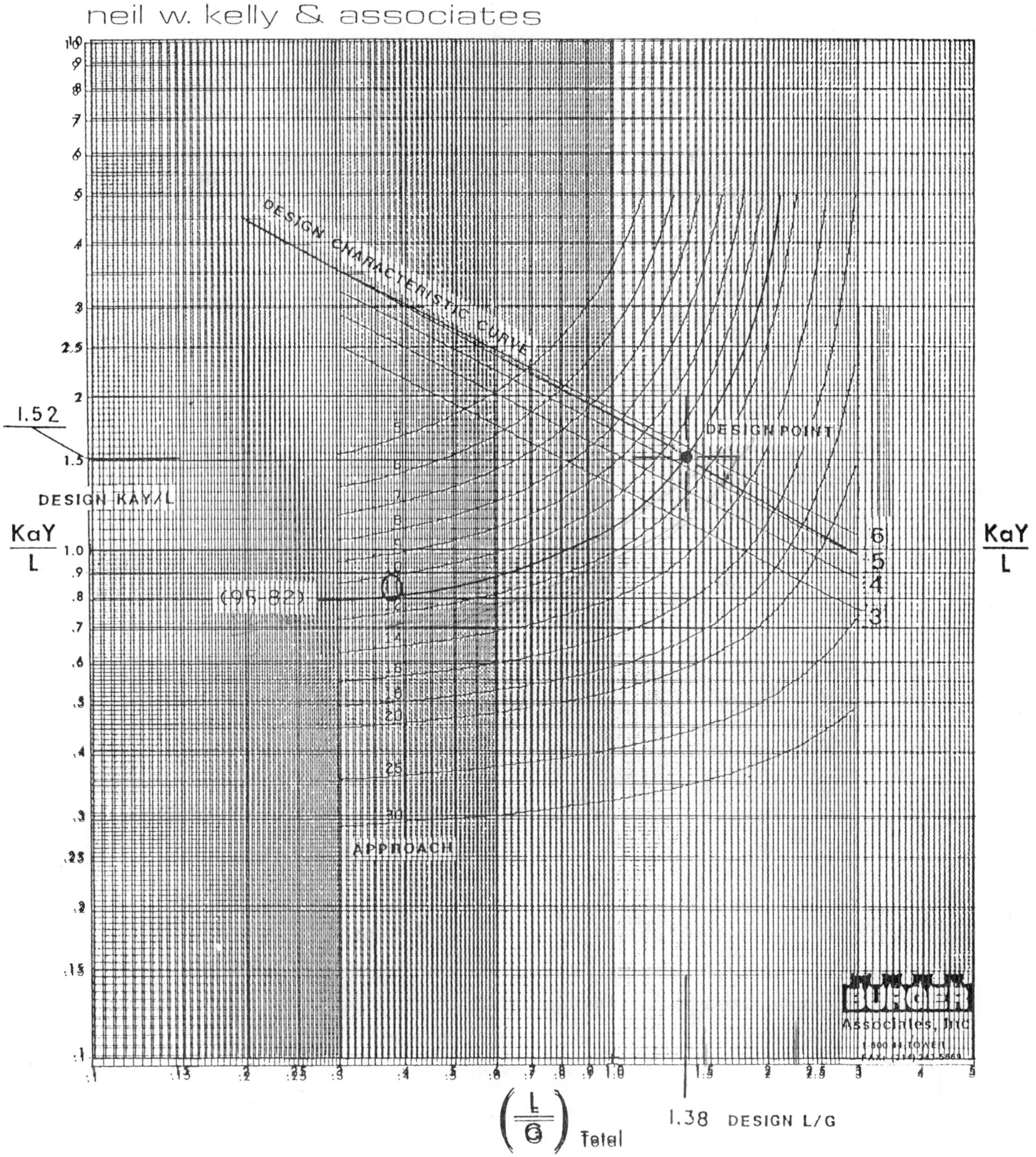

Figure 7

Step Four

Superimpose the performance curve transparency on top of the Design Point, aligning the left and right horizontal KaY/L 1.0 line and the vertical L/G 1.0 value to select the proper depth of fill which will be the value below the Design Point that will produce the calculated Design Conditions. Figure 7. Accordingly, fill manufacturers will provide static pressure losses based upon air velocity, water loading, and depth of fill which must be considered while selecting the depth of proper cellular fill materials.

In this exercise, 5½ feet of horizontal air travel through the 1900 or 19060 material will produce the Design Conditions.

Example:

II. Crossflow tower with a greater than 20' vertical fill height.

Design Conditions

GPM: 104,000
HWT: 120°F
CWT: 90°F
WBT: 82°F
CFM: 1,476,000 Per Fan

30°F Range
8°F Approach

8 Cells (13,000 GPM Per Cell)
28 foot diameter fans
200 HP Motors

$$\text{Design } \ ^{L}/_{G} = \frac{\frac{104{,}000}{8}}{1{,}476{,}000} \times 116.5 = 1.02$$

Testing indicated an 87.2% of capability, therefore, derate the air by 12.8% (1,476,000) (.872) = 1,287,072 CFM since the manufactured probably erred in his calculations.

$$\text{Actual } \ ^{L}/_{G} = \frac{\frac{104{,}000}{8}}{1{,}287{,}072} \times 116.5 = 1.18$$

Step One

The initial inquiry will be to locate the Design Point on a particular performance curve for these Design Conditions. Figure 8. Opening up the Kelly's Handbook of crossflow cooling tower performance curves, go to the 82°F wet bulb section and continue up to the 30°F range page. The Approach to the wet bulb (CWT of 90°F minus WBT of 82°F) is 8°F. With a french curve, darken the 8°F Approach parameter curve to give us a visual base line.

Step Two

Calculate the liquid to gas ratio as described on Page 5 which is 1.18. Traveling vertically on the 1.18 line, intercept the 8°F Approach at the Design Point. Figure 8. The design characteristic curve is a performance value that is furnished by fill manufacturers, and is in the area of -.6 negative slope. In the plain language, approximately a 60° angle. This is drawn in and used for testing purposes. Locate the Design Point and move horizontally to the left hand KaV/L value of 2.85. Bearing in mind that the Design Point is the intersection of these three items - L/G, Approach, and KaY/L.

Step Three

Select the fill manufacturer's curve that is likely to produce the results required. Figure 9. On these large towers, we have the option of many manufacturers' configurations of wood and plastic extruded forms. Rather than select fill on the basis of the manufacturer's promise, "trust me", we prefer installing materials that have been tested, are reliable, and the curves furnished by the manufacturer are accurate on the conservative side. All reputable manufacturers will furnish performance curves, which can be turned into transparencies on a xerox machine. In the case of this type of tower, we have had them tested after conversion with the "V" bars and the heat rejection was produced in accordance with these calculations and manufacturer's curve.

82 °F, WET BULB
30 °F, RANGE

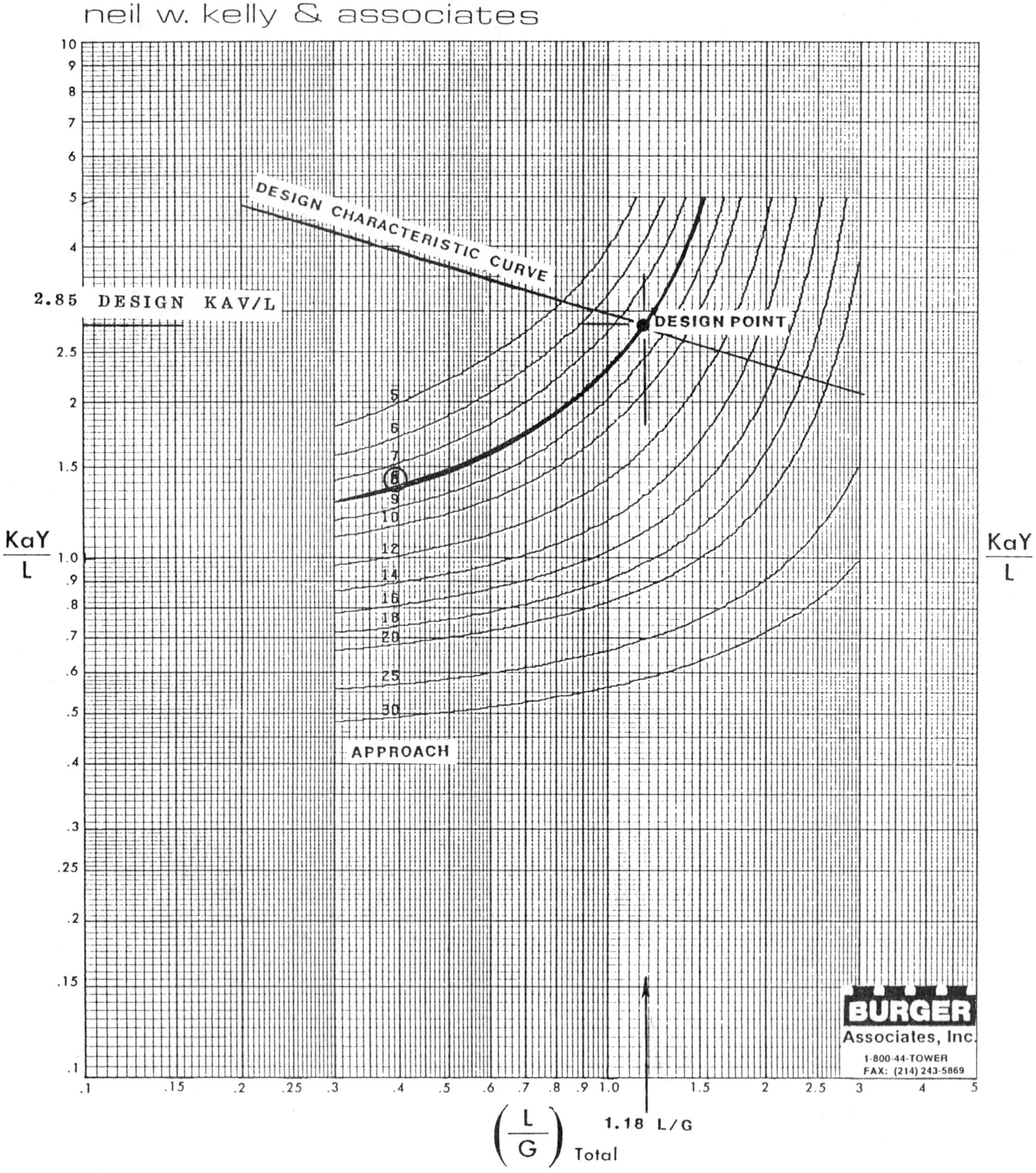

Figure 8

FILL G

MODIFIED

AIR TRAVEL = 20 FEET

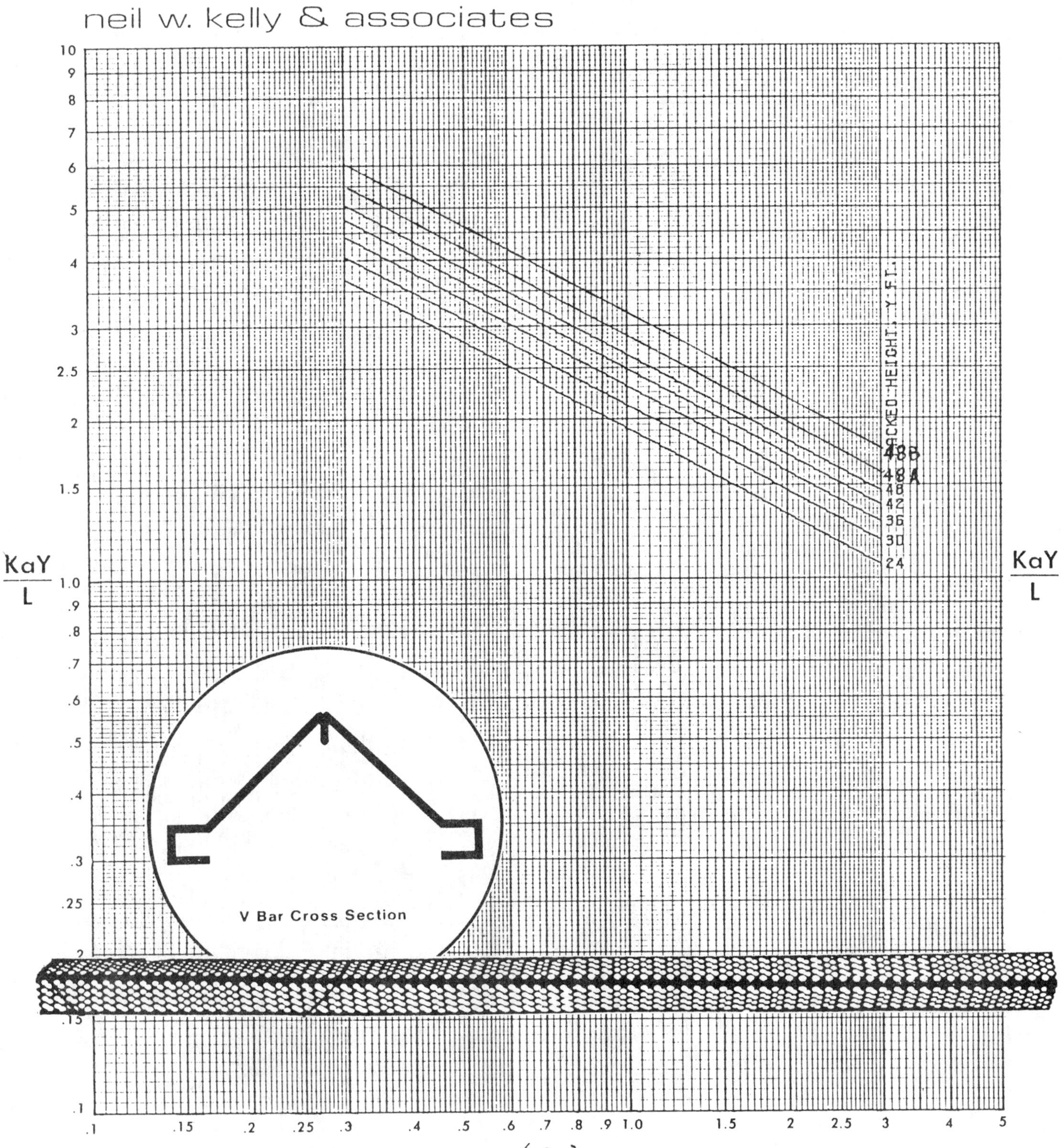

Figure 9

A = Fill Bars Parallel to Louvers

B = Fill Bars Parallel to Air Travel

AIR TRAVEL = 20 FEET

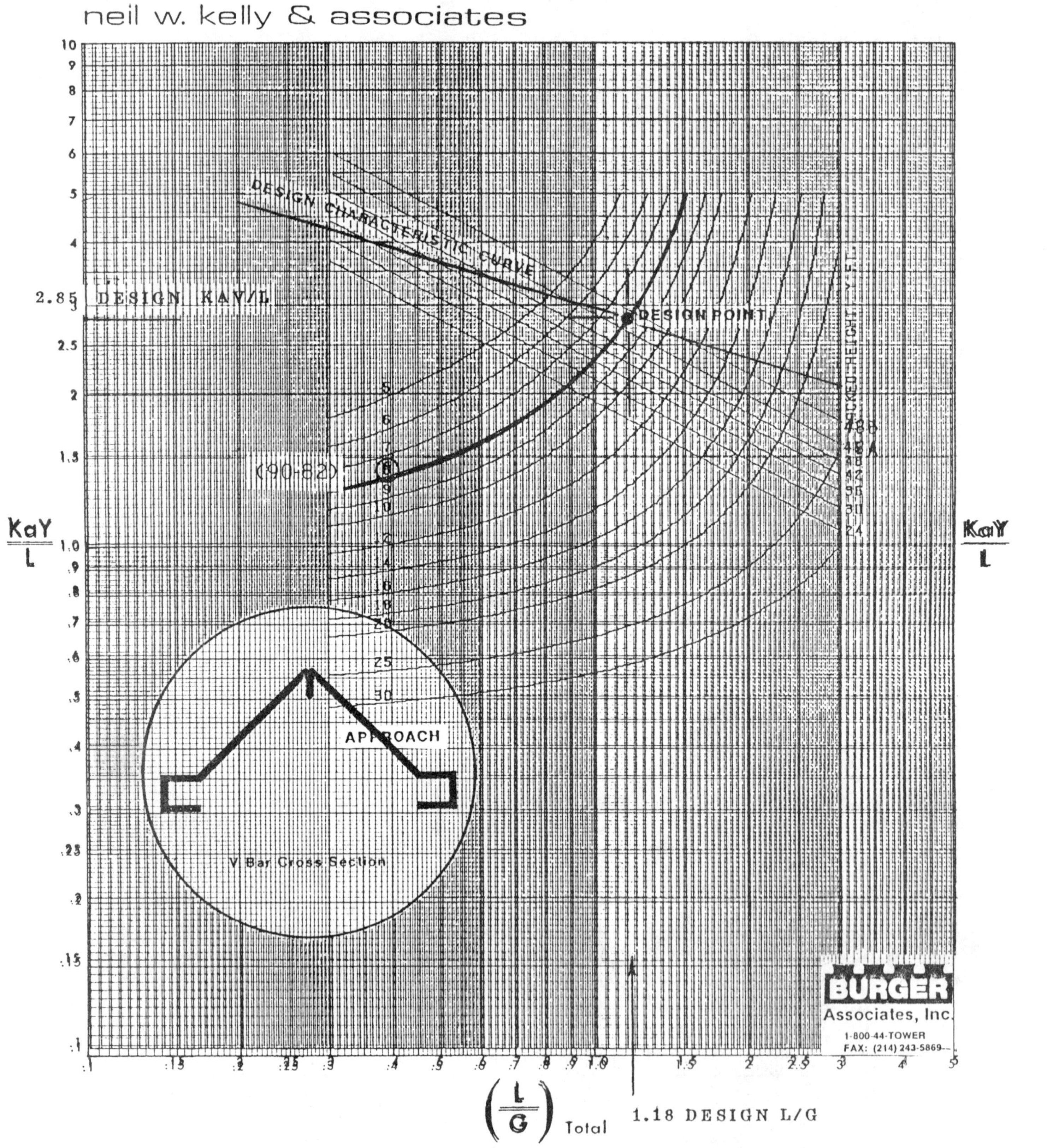

Figure 10

A = Fill Bars Parallel to Louvers

B = Fill Bars Parallel to Air Travel

Step Four

The Kelly Handbook illustrates performance curves for "V" bars in various configurations. Select the performance curve based upon air travel. In this case, fill G with an air travel of 20', since the tower air travel is fixed at 20 feet, meets the conditions of the cooling tower under investigation in this example. Superimposing the performance curve transparency, Figure 9, on top of the Design Point, it can be determined whether or not this material can be utilized. Figure 10. The right hand margin values on the fill "G" curves indicate the vertical height of the fill, which is a fixed value for each tower. Observe that 48B is above the Design Point indicating that this condition will produce Design Cooling. The Kelly analysis of the "V" bar fill is based on 8" vertical by 8" horizontal spacing of the fill bars installed at right angles to the air flow. By installing the bars parallel to the air flow, less static pressure is generated, and a film fill cooling effect plus splash cooling, increases the performance of the material. This is indicated by curve 48A. 48B is the same placement of bars parallel to the air flow but on 4" vertical by 8" horizontal pitch (spacing).

Each bar configuration must be tested independently since no two cross sections perform in the same manner. Some bars with flat horizontal surfaces and a right angle center piece are designed to be installed at right angles to the air so that the air turbulates against the vertical section. By installing this bar parallel to the air, the turbulating affect is lost and the performance is down-graded. The important criteria of fill selection is the manufacturer's integrity, track record, and testing to produce realistic performance curves to be used in rebuilding situations.

COUNTERFLOW COMPUTER PROGRAM

Today's selection science has proceeded from actual field testing ratings of unknown results to laborious hand working of the Tchebycheff Integration Formula utilizing enthalpy charts to Kelly and manufacturers logarithmic curves, to today's computer procedure. Each step upward refines the selection process and obtains a greater accuracy.

A computer program, especially for cooling tower technology, is only as good as the programmer's experience and expertise. Before purchasing and relying on a program for your calculations, it is advisable to obtain a sample run, build in accordance with the model, and then test the results to see if it is satisfactory.

The best type of program is one where all the known factors can be punched in and variables such as cold water temperature, water flow, range, L/G, and KaV/L values can be requested by the operator for the program to compute.

Then these variables can be changed until the desired tower capability of 100% or more is indicated.

The rebuilding example we will look at has the following parameters:-

GPM:	6,000	
HWT:	95°F	10°F Range
CWT:	85°F	6°F Approach
WBT:	79°F	
CFM:	300,000	Per Fan

13 foot diameter fans
40 HP Motor

The problem would be to select the type and amount of fill to produce a 100% tower.

The solution is to simply put the known quantities into the computer program with the fill depth at 0 feet and enter the different types of fill performance into the computer until a 100% or more tower is indicated. The solution will be printed out as to the required feet in the case 3.5 of the 1900 type material. Figure 11.

With this information, the next program is activated to produce curves at 110%, 100%, and 90% of water volume plotting wet bulb temperature against cold water temperature to determine the cold water temperature as the wet bulb changes.

Figure 12 illustrates one set of curves for the problem indicating at 79°F design wet bulb, the design point will produce 85°F water at the 10°F Range. The curves will also predict CWT for 8°F and 12°F Ranges.

Conversely when the wet bulb drops to 65°F the projected cold water temperature will be 74.25°F at the 10°F Design Range.

****

* I N D U C E D   D R A F T   C O U N T E R F L O W *

* BURGER & ASSOCIATES, INC. *

****

DALLAS COUNTY RECORDS BUILDING

T O W E R   S I Z E .

========================

CF1900

AIR INLET HEIGHT	ft	6.00	NUMBER OF CELLS		2
NUMBER OF FANS PER CELL		1	FILL DEPTH	ft	3.50
CELL LENGTH	ft	30.00	CELL WIDTH	ft	24.00
FAN DIAMETER..........	ft	13.00	HUB DIAMETER.............	ft	2.00
FAN EFFICIENCY........	%	70.00	TRANSMISSION EFFICIENCY...	%	96.00
VENTURI HEIGHT........	ft	4.00	STACK HEIGHT.............	ft	6.00
% BY-PASS ON WALLS.....	%	2.50	MOTOR SIZE (nominal)	HP	40.00
LOUVER P.D. COEFF......	VP	2.50	D E's P.D. COEFF........	VP	1.00
FAN INLET P.D. COEFF...	VP	0.20			
COEFF. OBST. INLET.....	VP	12.50	COEFF. OBST. FILL........	VP	2.00

T H E R M A L   D A T A .

============================

AMBIENT WET BULB	F	79.00	AIR REL. HUMIDITY.........	%	100
AIR DRY BULB...........	F	79.00	DESIGN WATERFLOW.......	GPM	6000
RANGE..................	F	10.00	DESIGN COLD WATER........	F	85.00
BAROMETRIC PRESSURE.	in Hg	29.39	ALTITUDE ABOVE MSL.........		500
KaV/L multiplying factor.		1.000	Fill Pr.drop multip. factor		1.000
MERKEL CORRECTION FACTOR		1.070	Recirculation allowance	F	0.00

R E S U L T S .

================

ACTUAL WATERFLOW	GPM	6031	TOWER CAPABILITY	%	100.51
WATERLOADING	GPM/ft2	4.2	Actual HP per fan	HP	36.0
FILL AIR VELOCITY		413	FAN AIR VELOCITY	FPM	2285
GAS/LIQUID G/L		0.816	LIQUID/GAS L/G		1.226

	AIRFLOW CFM	AREA ft2	VEL FPM	DENS lb/ft3	SP.VOL ft3/lb
INLET	572977	882	650	0.0713	14.33
FILL	582600	1411	413		
FAN	592224	259	2285	0.0696	14.81

F A N   P R E S S U R E .

============================

INLET 8.30E-7 * 2.50 * 0.0713 * 650 ^2 =..........	in WG	0.062
FILL as per manufacturer's data	in WG	0.125
DISTRIBUTION and PIPING included in fill data	in WG	0.000
DE'S 8.30E-7 * 1.00 * 0.0696 * 420 ^2 =..........	in WG	0.010
FAN VP 8.30E-7 * 0.0696 * 2285 ^2 =.................	in WG	0.302
FAN INLET 0.200* 0.302=	in WG	0.060
VENTURI STACK RECOVERY (if any)	in WG	-0.040
FAN STATIC PRESSURE (LOUVERS+FILL+DIST.+DE's.............	in WG	0.198
FAN STATIC PRESSURE (same as above + FAN INLET)...........	in WG	0.258
FAN STATIC PRESSURE (same as above - recovery	in WG	0.218
FAN TOTAL PRESSURE	in WG	0.520
AIRFLOW PER FAN..	CFM	296112
AIR EXIT WET BULB (for fan rating).........................	F	89.36

K A V / L

=========

WATER SPRAY ZONE (distribution)..................................	0.000
FILL...	1.491
RAIN ZONE (BELOW FILL)...	0.053
KAV/L TOTAL	1.544

Figure 11

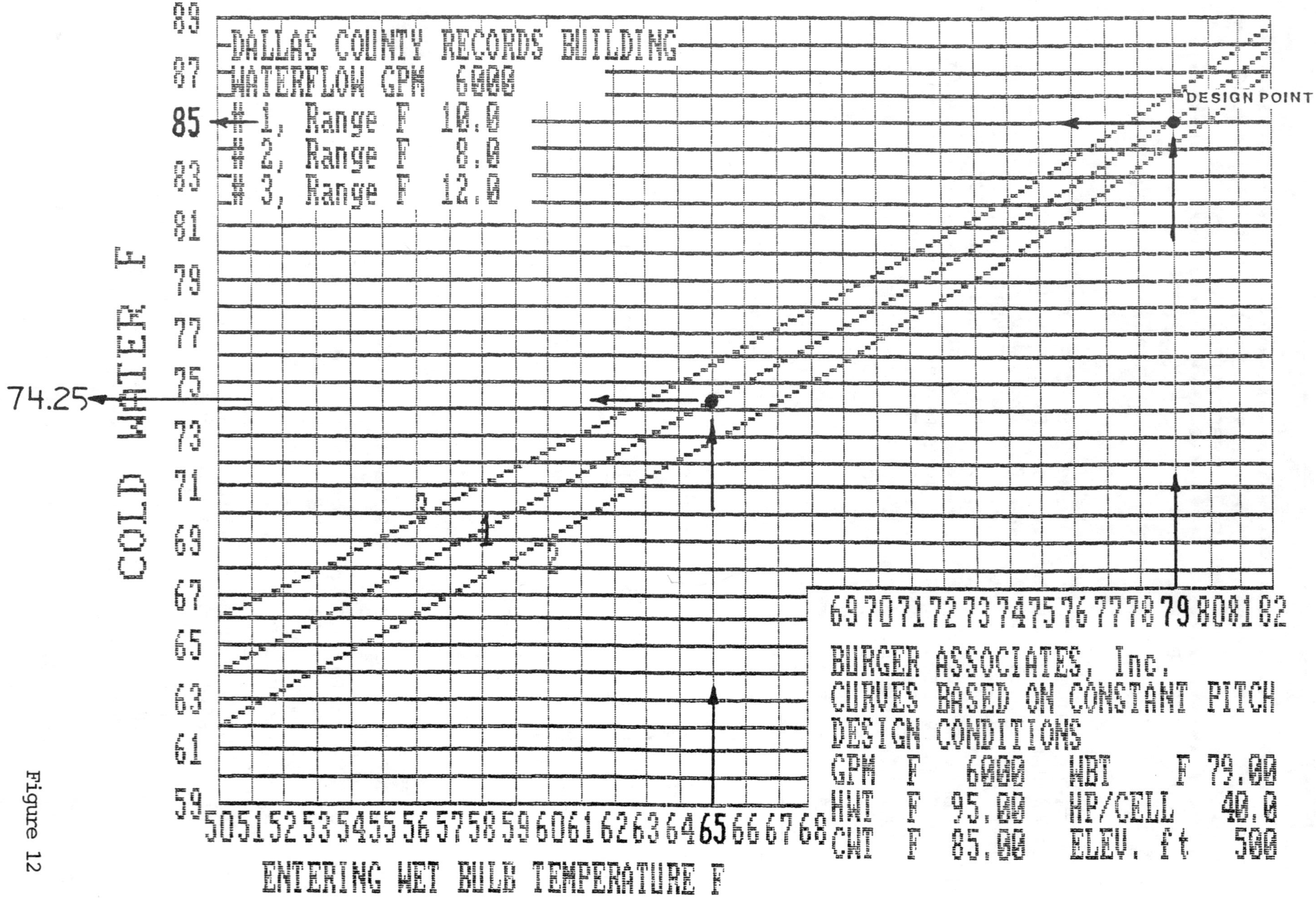
DALLAS COUNTY RECORDS BUILDING
WATERFLOW GPM 6000
1, Range F 10.0
2, Range F 8.0
3, Range F 12.0
DESIGN POINT
74.25
85
COLD WATER F
ENTERING WET BULB TEMPERATURE F
65
79
BURGER ASSOCIATES, Inc.
CURVES BASED ON CONSTANT PITCH
DESIGN CONDITIONS
GPM F 6000 WBT F 79.00
HWT F 95.00 HP/CELL 40.0
CWT F 85.00 ELEV. ft 500

Figure 12

TESTING YOUR TOWER

When a cooling tower is purchased or rebuilt, there are five variable factors in the Design Conditions that must be considered:

1. Circulating water volume (GPM)
2. Hot water temperature (HWT)
3. Cold water temperature (CWT)
4. Wet Bulb temperature (WBT)

    a. The Wet Bulb consists of two thermal measurements. Either the ambient temperature plus relative humidity or ambient temperature plus dew point.

Since these values constantly change, all five design conditions will never occur at the same time. For example, in the exercise on Page 8, 104,000 gallons of water per minute will never be circulated at that level. It will vary by plus or minus 10% or more when all pumps are operating. If one or more pumps are not operating, the volume could fall down to some 80,000 odd gallons per minute. By the same token, the 120°F design hot water temperature will never be 120.0°F but will range from 119.1°F to 122.5°F. If the process in the chemical plant or power plant is not working at 100% of capacity, that hot water temperature could be 105°F or lower.

Therefore, to determine a tower's performance, it must be tested with the test results then related to design requirements.

In the example on Page 52, the contractor or builder should be required to submit a Design Characteristic Curve and Design Point of his proposed cooling tower. Figure 8.

The five Design Condition points, in actuality, will probably never be achieved at one time because of the variable factors involved as explained above. Therefore the tower must be tested on "Off Design Conditions" and the results compared to Design.

TEST SUMMARY

TEST DATE 7-21-87

JOB # 6131

DESIGN CONDITIONS

GPM(US GALS) 104,000	DESIGN L/G 1.18	CWT °F 90
# CELLS TESTED 8	KAV-Y(L) 2.85	WBT °F 82
HP/ CELL 200	ACFM/CELL 1,287,000	RANGE °F 30
PUMP HEAD 55	HWT °F 120	APPROACH °F 8

TEST CONDITIONS

GPM (US GALS) 103,476	TEST L/G 1.45	CWT °F 88.7
# CELLS TESTED 8	KAV-Y(L) 2.05	WBT °F 76.2
HP/ CELL 186.2	ACFM/CELL 1,040,000	RANGE °F 22.4
PUMP HEAD 55	HWT °F 111.1	APPROACH °F 12.5

% CAPABILITY 84.7

Figure 13

76 °F, WET BULB
22 °F, RANGE (111.1 − 88.7)

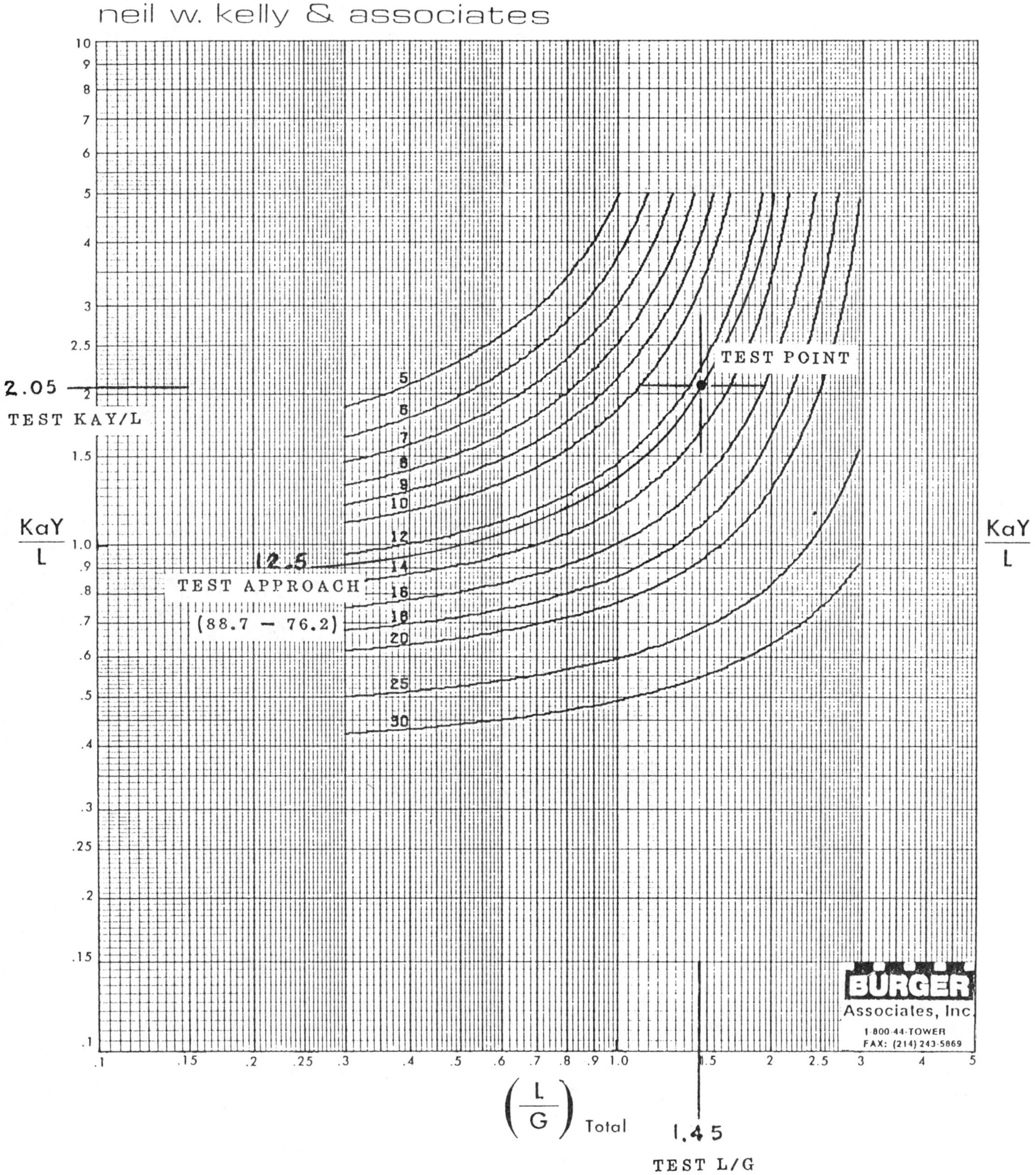

Figure 14

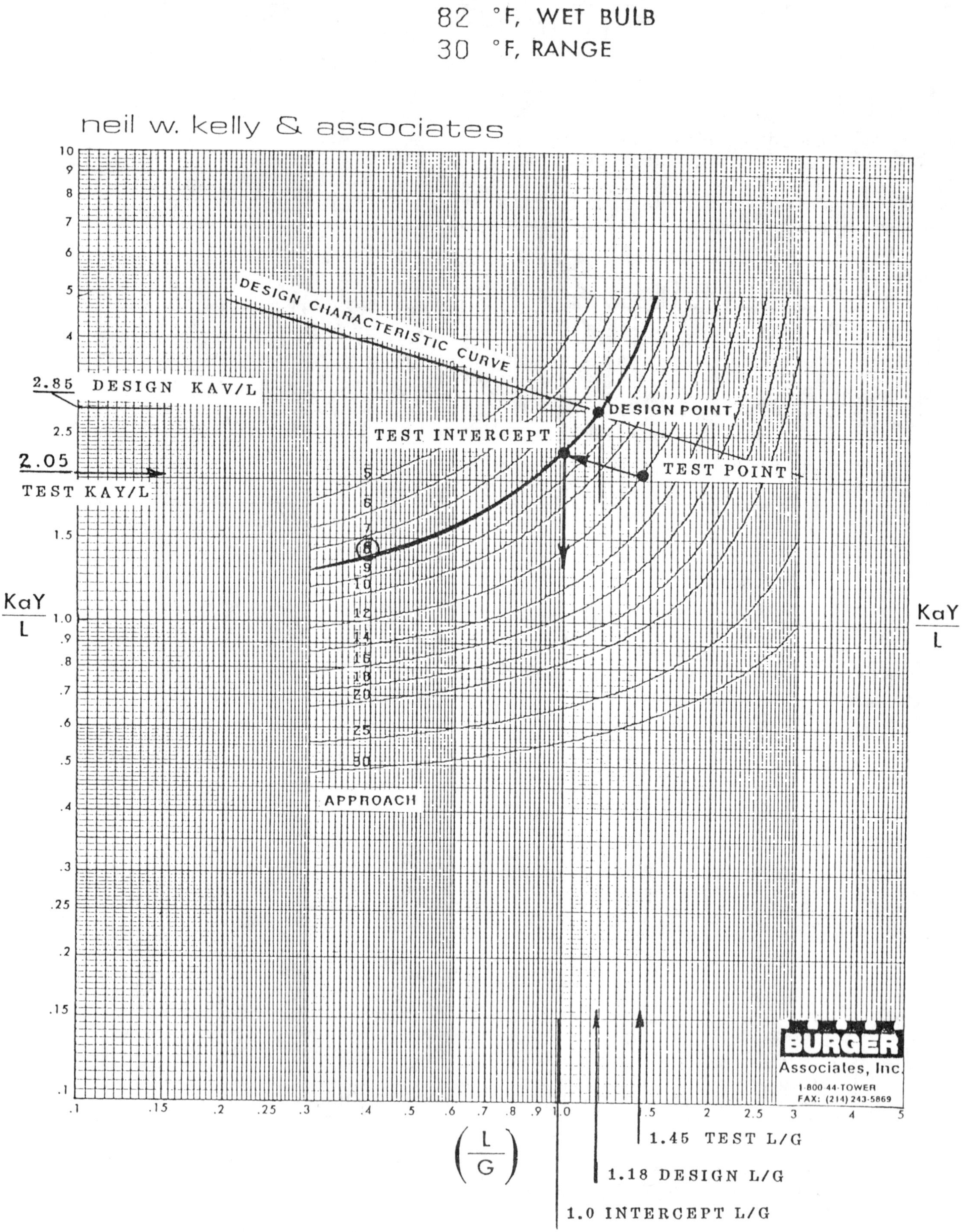
82 °F, WET BULB
30 °F, RANGE
neil w. kelly & associates
DESIGN CHARACTERISTIC CURVE
2.85 DESIGN KAV/L
DESIGN POINT
TEST INTERCEPT
2.05
TEST KAY/L
TEST POINT
APPROACH
KaY/L
(L/G)
1.45 TEST L/G
1.18 DESIGN L/G
1.0 INTERCEPT L/G
BURGER
Associates, Inc.
1-800-44-TOWER
FAX: (214) 243-5869

Figure 15

The test can be performed in accordance with the American Society of Mechanical Engineers (ASME) Standard PTC-23, or made by a manufacturer or contractor familiar with the procedure of using the testing equipment that is within the calibration periods stated in the test codes, or informally accomplished by operating personnel.

On the day of the test, parameters should be selected as close as possible to Design Wet Bulb, or within 10°F of Design WBT, water flow and heat load. If the test conditions are beyond the limitations of the Code, the results are too open to interpretation. The following hypothetical readings for the purpose of this illustration are measured every five minutes for the duration of one hour, resulting in this example, using the rebuilt tower described in Figure 8, of the Thermal Evaluation Chapter, of the following averages, Figure 13.

Hot Water Temperature	111.1°F
Cold Water Temperature	88.7°F
Wet Bulb Temperature	76.2°F
Water Flow	103,476 GPM
Air Flow	1,039,220 CFM
Range	22.4°F
Approach	12.5°F

These averages of actual operating conditions are compiled and a test point located on the performance curve of the Wet Bulb and cooling range values for the time of the test. The actual test field readings and resulting averages determine the Test Point. Figure 14.

According to the laws of logarithmic functions, since both Design Point and Test Point are plotted on the same logarithmic paper divisions, the relationships of the abscissa (L/G) X axis and ordinate (KaY/L) Y axis are the same at the different Wet Bulb and Range values. Therefore, we can transfer the Test Point from the test characteristic curve of 76°F WBT and 22°F Range, Figure 14, to the Design Characteristic performance curves of 82°F WBT and 30°F Range, Figure 15, at the 1.45 L/G and 2.05 KaY/L test values. We can then compare the theoretical Design Point to the actual Operating, or Test Point.

Draw a line parallel to the Design Characteristic curve, which is the performance of the particular fill in the tower being tested, to intercept the Design Approach. This is called the Test Intercept L/G. In this particular case, the value is 1.0.

$$\therefore \text{Capability} = Q = \frac{Q(\text{Test})}{Q\ (\text{Design})} \times 100 = \frac{1.0}{1.18} = 84.7\%\ \text{Tower}$$

In other words, the results of this theoretical exercise indicates that the tower tested was 15.3% short.

When the Test Point is transferred to the original Design Point curves, the new point (Test Point) is the actual real world Operating Point of the cooling tower and redefines the theoretical Design Point to the actual Operating Point. This point was physically tested and usually differs from the Design Point, sometimes optimistically calculated for sales hype.

CHAPTER 7

AIR MANAGEMENT

One of the significant factors for effective cooling is the control of air volume and flow through the structure. In natural draft towers, air movement is dependent upon the fluctuations created by the density difference between atmospheric air and air inside the tower. These humidity and temperature differenticals are beyond the regulation of the operator and will not be discussed in this chapter (fan assisted natural draft tower notwithstanding).

This section concerning air handling investigates methods by which the air can be more efficiently utilized to provide higher levels of heat transfer between the hot water and colder air.

An important factor in fan selection is the rating procedure to determine the proper fan diameter and size for the available existing horse power or the calculation of the proper motor horse power to be selected for maximum fan efficiency.

The procedures and information provided by the Hudson Products Corporation is greatfully acknowledged and the data is reprinted with their kind permission.

To aid the cooling tower Design Engineers and Operators major fan manufacturers will provide curves, technical data, and assistance to properly rate the performance of their air moving equipment.

By using the methods outlined on the following pages, the Engineer can investigate the required characteristics of a fan and driver. The rating procedure, sample problems, and fan curves are specifically from the Hudson Products Manual, which is available upon request by a user. This reference book consists of a family of curves for the available selection of fans.

This type of information can be obtained from other Manufacturers also. The nomograph chart procedures utilized by the Moore Fan Company is also illustrated which is fully explained in the Moore manual.

VARIABLE AIR FLOW CONTROLS

Three commonly accepted methods of energy-air control are described in this chapter. The basic concept is to monitor the cold water discharge basin with a "set point" that activates controls to regulate the air flow generated by the rotating fan.

Fan manufactures can provide an auto-variable pitch fan that automatically regulates the angularity of the fan blades, depending upon thermal demand, thereby reducing or increasing the air flow for the given horsepower.

Variable frequency alternators are used considerably in industry which changes alternating current to direct current, microprocessors determine the amount of D.C. required for the temperature needs, then reverses the current to A.C. at the lower value. This causes the fans to rotate at slower speeds, generating less air, or more air as the demand indicates.

The newest and most positive energy saving installation is the hydraulicly operated fan which eliminates the shaft and gear box from the top of the tower thereby reducing maintenance and hazards. The volume of hydraulic oil in the system is controlled, depending upon thermal demand. As the wet bulb drops, less volume of hydraulic fluid is in the system thereby rotating the fan at lower speed. When colder water is required, the process reverses, increasing hydraulic volume to speed fans up.

RATING PROCEDURE

Step 1 Determine Standard VELOCITY PRESSURE ($V.P_{Std.}$)

Using proper fan curve - read up from Design ACFM to V.P. reference line, then to right to read $V.P_{Std.}$

Step 2 Determine Actual TOTAL PRESSURE ($T.P_{Act.}$)

$$T.P_{Act.} = S.P_{Act.} + \left[V.P_{Std.} \div \frac{1}{\text{Density Ratio}} \right]$$

Read $\frac{1}{D.R.}$ from Chart I

Step 3 Determine Corrected TOTAL PRESSURE ($T.P_{Corr.}$)

$$T.P_{Corr.} = T.P_{Act.} \times \frac{1}{\text{Density Ratio}}$$

Step 4 Determine Required BLADE PITCH ANGLE

A. Curve available at Design RPM:
Enter proper fan curve at Design ACFM and $T.P_{Corr.}$ (Step 3).
Read Pitch directly from curve.

B. If fan is operating at RPM other than shown on curves:
Enter proper fan curve at Design ACFM X (S.F.) and $T.P_{Corr.} \times (S.F.)^2$
Read Pitch directly from curve.

Step 5 Determine Actual HORSEPOWER ($H.P_{Act.}$)

A. Curve available at Design RPM:
Enter curve at Design ACFM and move vertically to intersect Horsepower curve at pitch angle determined in Step 4A. Read $H.P_{Curve}$ on scale to left.

$$H.P_{Act.} = H.P_{Curve} \div \frac{1}{\text{Density Ratio}}$$

B. If fan is operating at RPM other than shown on curves:
Enter curve at Design ACFM X (S.F.) and move vertically to intersect Horsepower curve at Pitch angle determined in Step 4B. Read $H.P_{Curve}$ on scale to left.

$$H.P_{Act.} = H.P_{Curve} \div \left[\frac{1}{\text{Density Ratio}} \times (S.F.)^3 \right]$$

Step 6 Determine FAN EFFICIENCY

$$\text{Fan Eff.} = \frac{ACFM \times T.P_{Act.}}{6356 \times H.P_{Act.}}$$

FAN LAWS

$$CFM_{CURVE} = CFM_{ACTUAL} \times \left[\frac{RPM_{CURVE}}{RPM_{ACTUAL}} \right]$$

$$H.P. = H.P_{CURVE} \div \left[\frac{RPM_{CURVE}}{RPM_{ACTUAL}} \right]^3$$

$$T.P_{CURVE} = T.P_{CORR.} \times \left[\frac{RPM_{CURVE}}{RPM_{ACTUAL}} \right]^2$$

$$\text{SPEED FACTOR} = \frac{RPM_{CURVE}}{RPM_{ACTUAL}} = S.F.$$

SAMPLE PROBLEMS

A. 12 ft. Dia. Fan
RPM = 318
ACFM = 200,000
Air Temp. = 170°F
$S.P._{Act.}$ = .491 in. H_2O
Alt. = Sea Level

Step 1
Read up from 200,000 to V.P. line and to right to .235 in. H_2O

Step 2
$$T.P._{Act.} = .491 + \frac{.235}{1.19*} = .688$$
*Read from Chart I (Page 3)

Step 3
$$T.P._{Corr.} = .688 \times 1.19 = .82$$

Step 4A
Enter T-12B-4 Curve at 200,000 ACFM and .82 T.P., Read 14° Pitch

Step 5A
$H.P._{Curve}$ = 31.0 (Read Directly from curve)

$$H.P._{Act.} = \frac{31.0}{1.19} = 26.05$$

Step 6
$$\text{Fan Eff.} = \frac{200{,}000 \times .688}{6356 \times 26.05} = .83$$

B. Same as above except -
RPM = 292 Therefore S.F. = $\frac{318}{292}$ = 1.089

Step 1 through Step 3 same as above.

Step 4B
Enter T-12B-4 curve at 200,000 X 1.089 = 217,800 ACFM and .82 X $(1.089)^2$ = .972. Read 18° Pitch.

Step 5B
$H.P._{Curve}$ = 42.5 (Read directly from curve)
$$H.P._{Act.} = 42.5 \div \left[1.19 \times (1.089)^3\right] = 27.7$$

Step 6
$$\text{Fan Eff.} = \frac{200{,}000 \times .688}{6356 \times 27.7} = .78$$

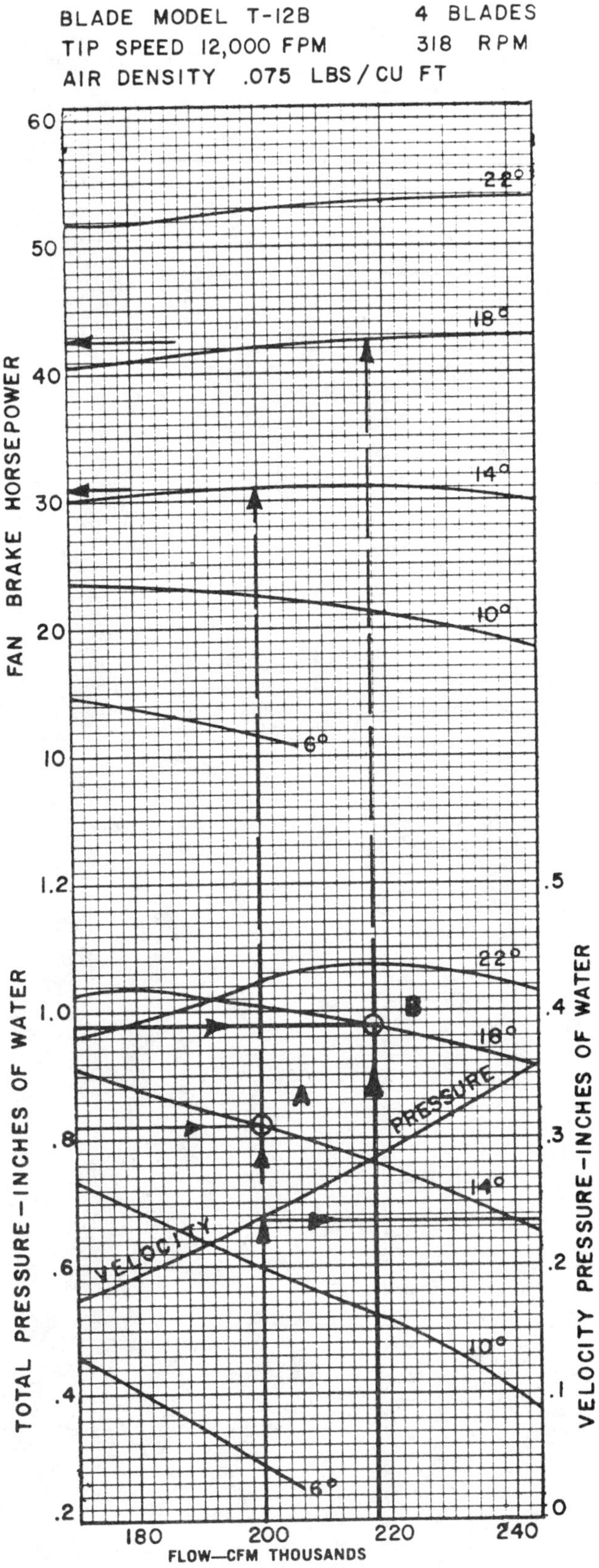

CHART NO. 1

NOMOGRAPH FOR DETERMINING AIR DENSITY CORRECTION

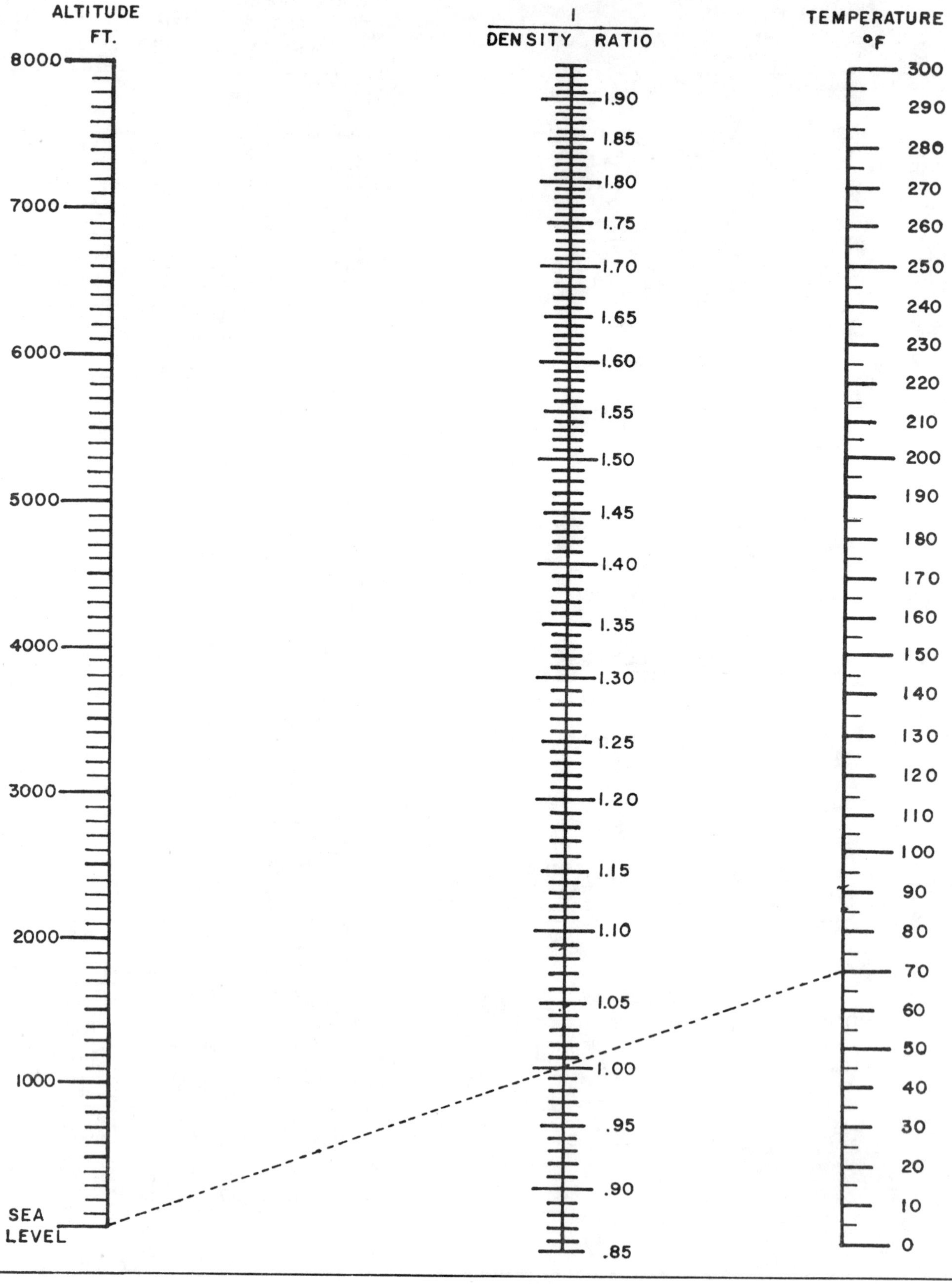

SELECTING CLASS 5000 BLOWERS

USE OF CHARTS

THE Moore COMPANY

TEN SIMPLE STEPS

Only ten simple steps are necessary to select the right Class 5000 Blower from hundreds of thousands of possible combinations. Nine of these steps consist of specially prepared nomographs which are read with a straight edge. The tenth step is a simple addition made in space provided.

The chart at right is for a Series 40 unit. Similar charts are provided for Series 17, Series 27, Series 60 and Series 90 units on the following pages. These charts are available in pad form on request.

SELECTION EXAMPLE

A simple selection will be carried through each step and illustrated by dotted lines on the charts on the opposite page. For this example, let us assume the following performance requirements:

Density (or temperature and elevation):	100° F., 1000 Ft.
Cubic Feet Per Minute:	145,000 SCFM
Static Pressure:	.45" H_2O
Blower Diameter:	12 Feet

1 DENSITY RATIO

If the density of the air being handled by the blower is given in pounds per cubic foot, the density ratio can be found opposite this figure on the Density Ratio Chart. If the temperature and elevation are known, connect the proper temperature on the temperature scale with the elevation on the elevation scale and read the density ratio from the center scale where it is intersected by this connecting line.

EXAMPLE:

Given 100° F. at 1000 Ft. elevation. On the Density Ratio Chart, connect these points and read a density ratio of .91. (Note that this corresponds to a density of about .068 Lbs. per Cu. Ft. It is density ratio that is to be used in fan selection, not density.)

2 ACTUAL CUBIC FEET PER MINUTE

The Cubic Feet Per Minute (CFM) will be given as either Actual Cubic Feet per Minute (ACFM) at the proper density ratio or in terms of standard air (70° and sea level). If CFM is given in terms of standard air (SCFM), it must be converted to the actual CFM to be handled by the fan. This is easily accomplished by connecting the proper density ratio with the standard CFM desired and reading the actual CFM on the center scale. Blowers must always be rated for actual cubic feet per minute (ACFM) at the proper density ratio.

EXAMPLE:

Given 145,000 SCFM required, connect this point to .91 on the density ratio scale on the ACFM Chart and read 160,000 ACFM on the center scale.

3 VELOCITY OR BLOWER DIAMETER

With a straight edge on the ACFM required on the top scale of the Velocity Chart, select the diameter desired and extend a straight line to the velocity scale on the bottom. This will give the velocity of the required CFM through a unit of a given diameter, which should not exceed 2500 FPM unless a bell inlet is provided.

EXAMPLE:

Since a 12-Ft. unit is desired, we extend a straight line from 160,000 on the ACFM scale through 12-Ft. diameter on the center scale and read a velocity of 1500 FPM on the bottom scale

4 VELOCITY PRESSURE

Velocity pressure may be obtained from the Velocity Pressure Chart by connecting the proper points on the upper and lower scales and reading the velocity pressure on the center scale.

EXAMPLE:

For a velocity of 1500 FPM and the density ratio of .91, the velocity pressure will be found to be .13" H_2O.

5 TOTAL PRESSURE

To obtain total pressure, add the velocity pressure obtained above to the static pressure against which the blower is to operate.

EXAMPLE:

The given static pressure of .45" added to the velocity pressure of .13" gives a total pressure of .58" H_2O.

6 HORSEPOWER

In order to obtain the horsepower input to the fan, connect the proper points on the ACFM and the total pressure scales on the Horsepower Chart and read the brake horsepower input at 75% efficiency on the center scale. (If the driver to be used is of greater horsepower than that calculated to be consumed by the fan, the fan should be selected to consume the full horsepower of the driver. See Page 9.

EXAMPLE:

Connect .58 on the total pressure scale with 160,000 on the ACFM scale and read 19.0 BHP on the center scale.

7 PRESSURE PER BLADE

The maximum pressure which one blade of a given series will develop at a specified RPM may be obtained by connecting the desired RPM with the proper density ratio and reading the pressure per blade on the Pressure per Blade Chart.

EXAMPLE:

Try 260 RPM. Connect .91 on the upper scale to 260 on the lower scale and read a pressure per blade of .098" at 100% blade load.

8 NUMBER OF BLADES

Connect the proper point on the total pressure scale to the pressure per blade as determined above and read the number of blades required. This fractional number of blades will be used below to determine deflection. The next larger integral number of blades will, of course, actually be installed on the fan.

(NOTE: On new installations or under any circumstances where performance conditions may vary from the anticipated, it is well to select a number of blades equal to 115% to 125% of those indicated by the chart.)

EXAMPLE:

Extending a straight line from .58 on the total pressure scale to .098 on the pressure per blade scale, read 5.9 blades on the center scale. (Six blades would perform the stated requirements, but an additional blade may be desirable for questionable applications as explained in paragraph above.

9 PITCH IN HUB DIAMETERS

In order to determine the deflection at the hub, it is necessary to find the pitch at which the unit is operating. Connect the proper point on the R PM scale of the Pitch Chart to the point on the velocity scale representing the velocity already obtained, and read the pitch in hub diameters on the center scale. If possible, pitch should be maintained between 1-1/2 and 2-1/2 hub diameters - otherwise, see following pages.

EXAMPLE:

Connect 260 RPM with 1500 FPM and read a pitch of 1.72 hub diameters.

10 DEFLECTION AT HUB

On the deflection chart, connect the proper point on the pitch in hub diameters scale to the proper point on the number of blades scale and read the deflection at hub on the center scale. This angle will be the degrees of deflection from true axial flow and must not exceed 45° in fans without guide vanes.

EXAMPLE:

Connecting 1.72 on the pitch in hub diameters scale to 5.9 on the number of blades scale, we read a deflection of 45° on the center scale. This can be seen to be a satisfactory selection.

SELECTING CLASS 5000 BLOWERS

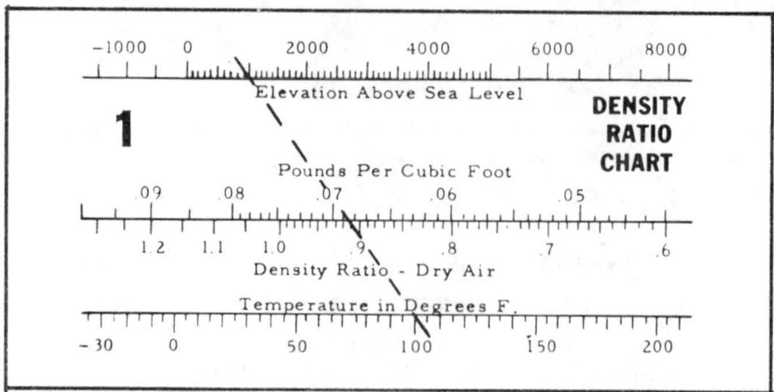

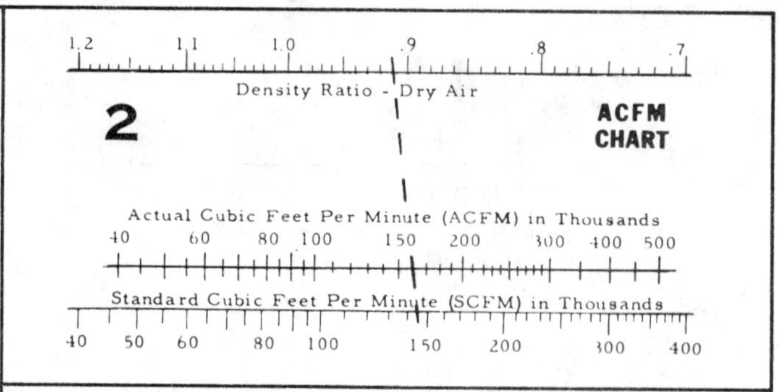

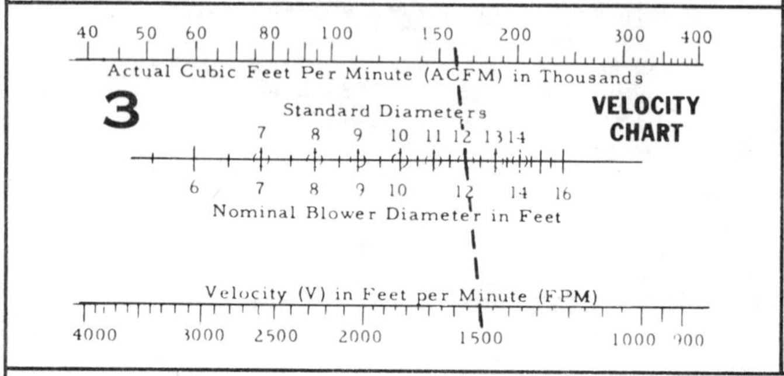

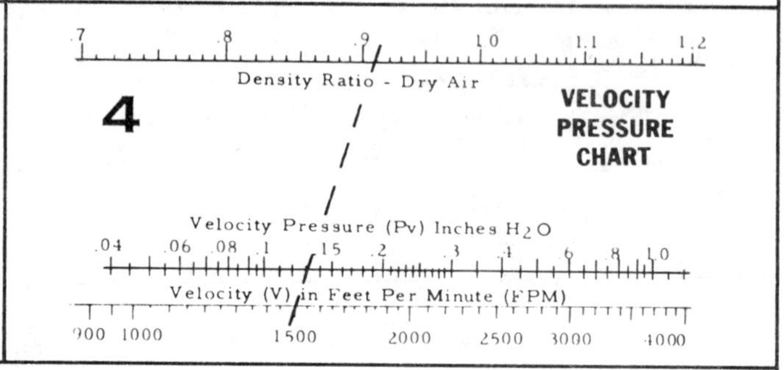

5 Static Pressure (Ps) 45″ + Velocity Pressure (Pv) .13″ = .58″ **TOTAL PRESSURE**

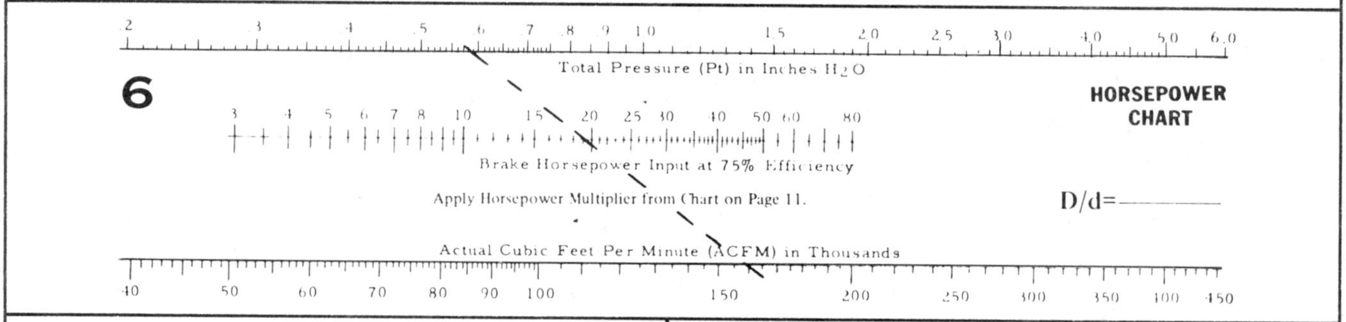

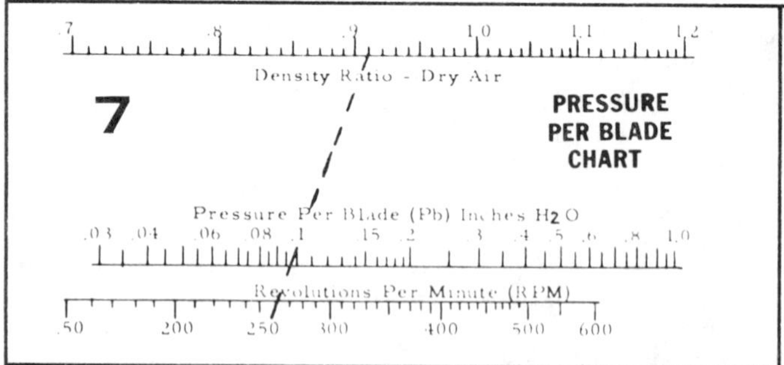

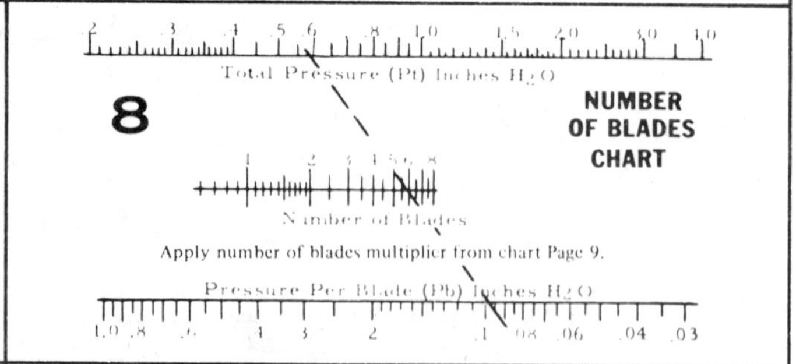

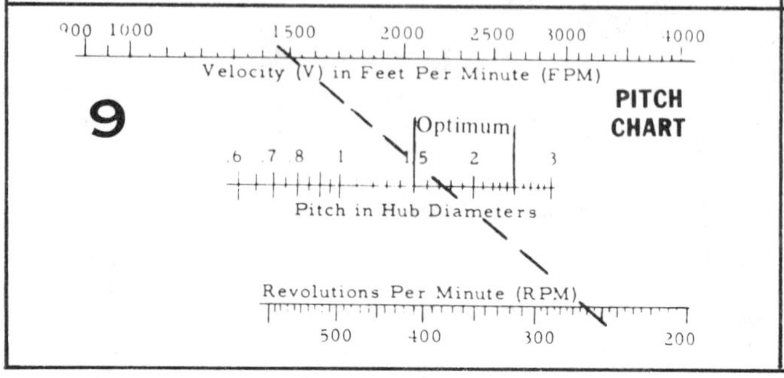

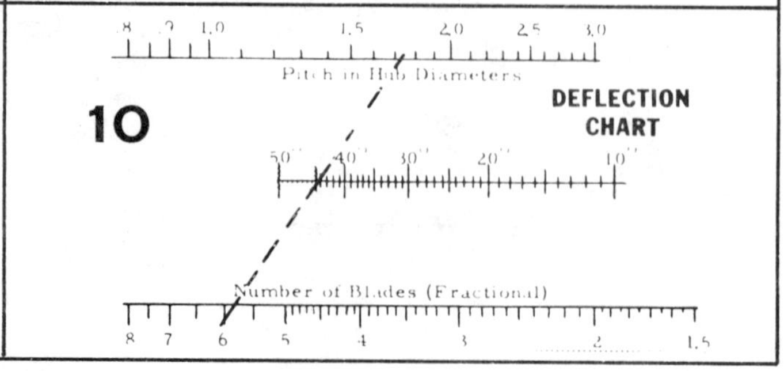

AUTO-VARIABLE PITCH FANS

1. Auto-variable pitch (AVP) fans are a viable option for energy conservation. Air movers and motors are sized by Designers for maximum required duty and maximum cooling to be applied at all times.

2. Since fans operate with constant speed and set pitch, airflow does not vary, and a constant electric load is drawn by the fan motors leading to excessive power consumption because maximum cooling is not needed 100% of the operating time.

3. Auto-variable pitch fans provide accurate outlet temperature control under conditions of varying heat load. They operate with minimum horse power at all times since the fans only move the amount of air actually needed. The system senses load automatically and adjusts to the most economical operating mode. During winter operations, the fans can be cycled to reverse air flow to prevent ice build up.

4. A cost-benefit analysis (enclosed) provides a payback period for retro-fit of two years for an electric cost of $.006/KWH. The higher the power cost, the faster the investment recovery. The variable pitch feature conserves a significant amount of power by its ability to modulate air flow as ambient temperature decreases.

5. The air supply requirements, to be furnished by others, are 3-15 psi instrument quality signal air, 25 psi supply air, and center support over fan.

New Motor Control Stabilizes Load Demand & Cuts Energy Costs

Variable Speed Drive

Figure 1 — Multi-cell refinery cooling tower totaling 600 HP driving 6 fans can conserve close to $100,000.00 or nearly 2,000,000 kilowatt hours per year when utilizing automatic variable frequency drive.

In these days of energy conservation awareness, equipment operators are besieged from all sides to retrofit in the name of power minimization.

Typical of power and energy wastage is a multi-cell cooling tower, Fig. 1, where motors and fans are designed to produce 100% of design cooling air, regardless of ambient conditions. If the cooling air can be controlled to produce constant design cold water regardless of the continually fluctuating wet bulb temperature, significant power, wear and tear, and maintenance savings can be obtained.

Since energy going into the motor and application is so closely controlled and regulated, there is no excess power wasted, which means that considerable energy savings can be realized.

A newcomer to the energy conservation club is the variable frequency drive (VFD) speed control of AC motors.

What is VFD?
Primarily, it is a method of controlling the speed of an AC squirrel cage induction motor, the workhorse of American industry, permitting it to perform a function at optimum efficiency at changing load conditions.

What are the benefits of VFD? Are the savings calculatable? What is the return on investment, (ROI)?

Variable frequency (speed) control is a method of adjusting the speed of the motor and the driven equipment so as to continuously satisfy the variable load requirements and thereby consume minimum energy. VFD also economically limits the amount of energy into the AC motor, permitting it to perform at a predetermined level of demand for optimum efficiency of operation.

Microprocessors, (MP's) and programmable controllers, (PC's) that dramatically alter drive system characteristics are a key break-through in drive control. MP's and PC's make it possible to use on-line mathematical expressions for motor control and process characteristics; to communicate with process or management computers as part of an overall control scheme; and to run tests and diagnostic functions automatically or when problems arise. Process control, sequencing, alternation, and various alarm functions are provided using microprocessor control which also sets "tamper-proof" limits that cannot be bypassed by over-zealous operators.

If a malfunction alarm is ignored, and conditions continue to deteriorate, the microprocessor then issues a shutdown command. Shut-down occurs in a safe, orderly sequence and a restart is now allowed until the fault condition is corrected.[2]

Demand Billing

Many utility companies are invoicing customers based upon peak kilowatt demand. Since motor current is under constant control utilizing VFD, starting inrush never exceeds 100% of full load current, yet full load motor torque is available for starting. This eliminates the demand spikes as high as 600% on some compressor, air conditioning, or fan motor start-ups. Utilizing VFD, utility bills in these areas can be considerably reduced.

COOLING TOWER APPLICATIONS
Cooling tower motors and fans are engineered to produce design conditions of 100% CFM all the time. However, by definition, design wet bulb does not occur but 2½% of the time, on the average, during the warmest consecutive 4 months as determined by the mean WBT. (3). Therefore 97½% of the time, the tower is over-producing on air and cold water production thereby wasting power consumption, i.e., during fall, winter, spring, and evenings.

To analyze the energy savings available requiring a multicell cooling tower to produce constant design cold water temperature (CWT), the past two or three years of wet bulb temperatures are averaged. These reports are available from the U.S. Weather Bureau Climatological reports. Fig. 6. Daytime and nighttime averages are separated since there will be significant differences in the averages. The monthly AM and PM values are then plotted from a design wet bulb fan power curve. This curve is the net fan horsepower (FHP) or KW plot corrected for the system efficiency of the fan unit and motor and can be used directly to define required power for any WBT condition.

To obtain a constant cold water temperature (CWT) with a varying wet bulb temperature (WBT) a reference must be established for the analysis of power consumption. Fig. 2 demonstrates the relationship of WBT variation with CFM, RPM, and Fan HP when CWT, HWT, and water flow rates are held constant.

The fan horsepower (FHP) curve is the plotted relationship of CFM to FHP which is a cubed variation with CFM.

The CFM and RPM Curve Fig. 3, is the plot of changing air flow requirements with fluctuation in WBT. This curve reveals the manner in which cooling tower performance is influenced by the changing environment and is mathematically derived utilizing basic cooling tower thermal computations (5) in conjunction with the psychrometric chart (6).

For the example in this text, we analyze a Gulf Coast Refinery process cooling tower[4], Fig. 1 having 6 cells with 100 horsepower electric drivers each to cool 42,000 GPM of circulating water entering at 120°F leaving at 90°F temperature. Fig. 4 is a plot of 100% power design point at 80°F WBT and interpolating the actual past three years night and day wet bulb monthly averages calculates the percent of design fan horsepower required on a monthly basis to produce constant design cold water temperature, Fig. 5 The 480 total horsepower requirements of the 6 cells, if operating at 8760 available design hours during the year would require $217,832.00 of electric usage at the average 5¢ per Kwh in the area. Fig. 6 indicates, mode 2.0 which reflects a practical design hours utilization operating at 85% of the time. Mode 3 is the actual design hours required to produce a constant 90°F CWT with automatic CFM control. This would result in $99,240.00 savings which divided into the net cost of installing on a turn-key basis, the temperature controllers, instrumentation, and inverter system package of $117,400.00 results in a .785 year or 9½ month ROI.

It is significant to note that the U.S. Public Law, P.L. 95.618 provides for a 10% tax credit to be allowed for energy conservation measures expended by industry, plus a 10% equipment investment for credit making the installation charge of $117,400.00 a net cost to the Company of $93,920.00.

If 100% of available design hours are used, than the payback relative to Mode 1.0 would be .621 years or an 8 month payback.

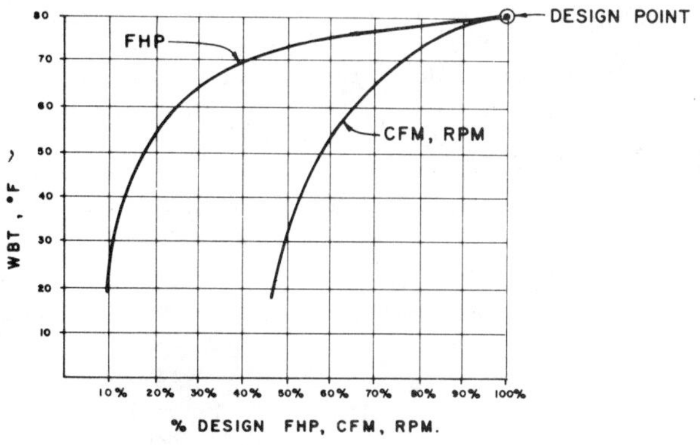

Fig. 2 — WBT-FHP characteristic.

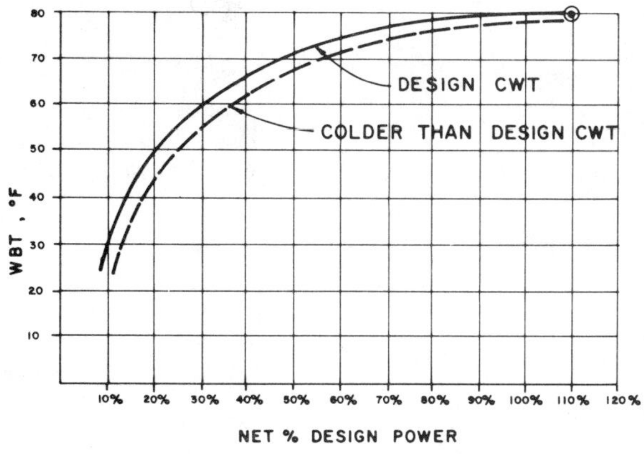

Fig. 3 — Net power.

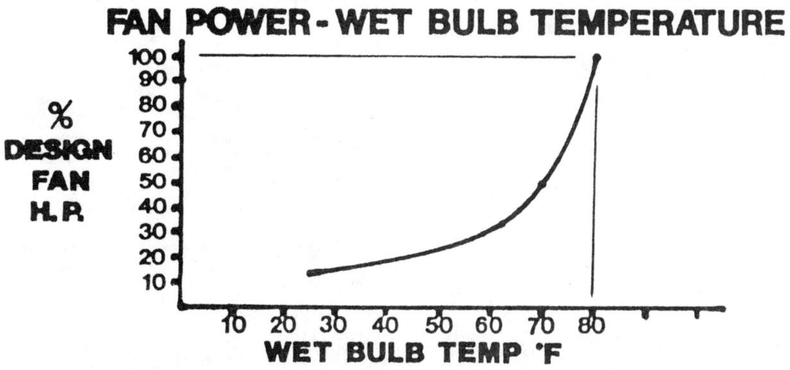

Fig. 4 — Wet bulb temperature vs. percent design horsepower that is required as the wet bulb varies.

The flexibility is unlimited. For example, a regulator cage with up to eight plug-in printed circuit boards, Fig. 7, can provide over two million combinations of functions. This makes it possible to configure drives plus specific applications and provide additional functions to other systems.

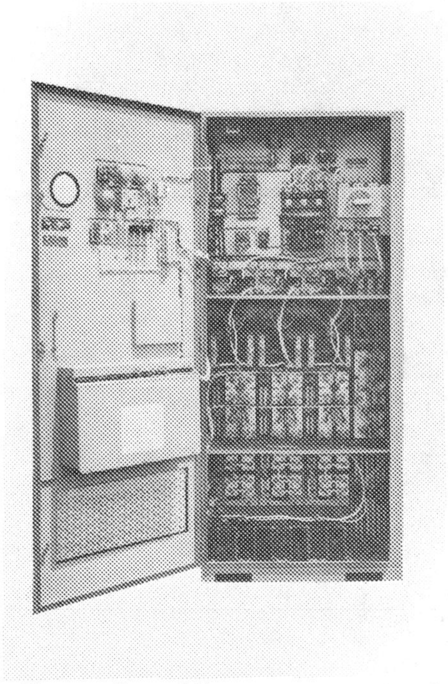

Figure 7 — Microprocessors and programmable controllers together with modern solid state power semi-conductors and integrated circuits sequence the functions to optimize energy conservation.

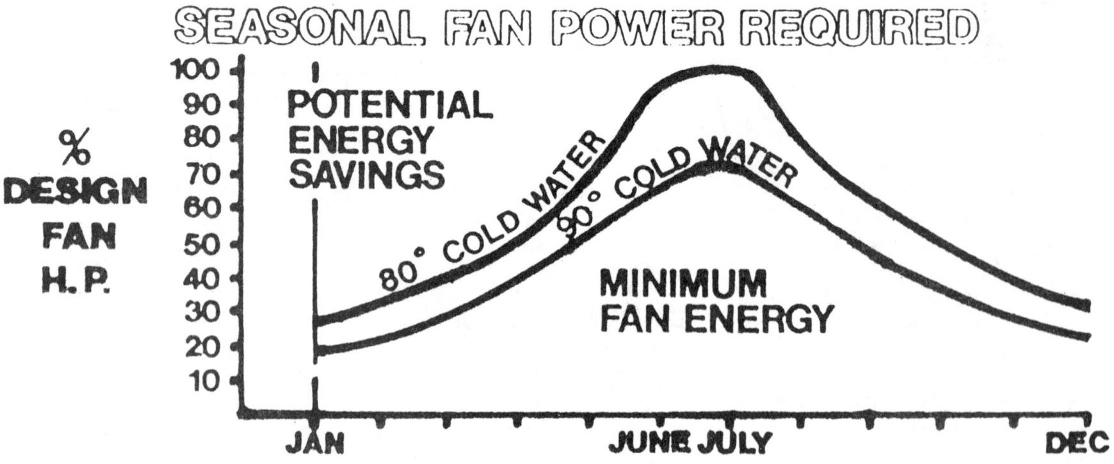

Fig. 5 — The relationship between design fan horsepower and the monthly fluctuation of the wet bulb.

Burger Associates, Inc.

ENERGY MANAGEMENT SYSTEMS

Job No. 5285

Date 7-21-84

COOLING TOWER ENERGY ANALYSIS

Requested by: RICHARD [illegible] (4)

Name, Title ENERGY CONSV. ENGR.

Company [illegible] CHEMICAL CO. INC

Address PO BOX [illegible]

HOUSTON, TEXAS

Phone: (713) [illegible]

DESIGN DATA – TOWER STYRENE

// Cells 6 HP/Fan 100 (600 TOTAL)

CWT 90°F GPM 42,000

HWT 120°F WBT 80°F

Yearly Average WBT Range 1979-80 44-77°F

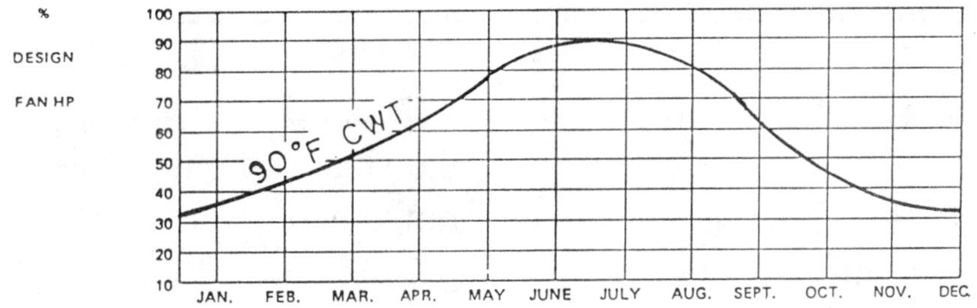

SUMMARY:

		Kw–Hrs.	Cost @ 5 ¢/Kw. Hr.
1.0 8760	Available Design Hours, 100%	4.36×10^6	$ 217,832.00
2.0 7446	Design Hours–ON/OFF Control, 85 % on 15 % off	3.70×10^6	$ 185,157.00
3.0 3839 CFM Control.	Design Hours @ 90°F CWT with Automatic	1.71×10^6	$ 85,917.00

4.0 Savings with Automatic CFM Control — SAVINGS:

Mode 1.0 – 3.0 $ 131,915.00

Mode 2.0 – 3.0 $ 99,240.00

5.0 Cost of Automatic CFM Control AF Unit and Instrumentation — COST: $ 117,400.00

6.0 Less 10% Credit P. L. 95.618 $ 11,740.00

Less 10% Investment Credit $ 11,740.00 — NET COST: $ 93,920.00

7.0 PAY BACK @ ¢ / Kw–Hr 2.0 .95 Months

(Referenced to Mode 1.0) .71 Months

Figure 6— The energy analysis sheet. (4)

CONCLUSION

The application of automatic speed control by the use of variable frequency AC drive is a new tool by which industry of all types can improve energy conservation and profits.

Based upon fluctuating wet bulb throughout the year versus constant 100% design air volume, the savings are quite dramatic with pay-backs in multi-cell towers from 9 months to a year and a half. Once the retrofit equipment, which does not interfere with constant plant operation, is installed, the energy savings can provide increased operating profit for the facility.

About 90% of the electric motors used in industry are three phase AC motors. Unfortunately, their speed is determined by their construction and by the frequency of the AC power. Normal plant power frequency of 50 or 60-Hertz is constant; hence the motor speed is constant. VFD is a modern control of motors whose application can be implemented without interruption of plant operation and without change in the rotating equipment. The turn-key package of controllers, sensors, instrumentation, and housing is installed — conveniently in the motor control center — or as close as practical to the equipment. Down time for each motor is held to a very small minimum to reconnect leads since the motor and it's rotating equipment is not altered.

To the directly computable electric energy savings must be added the increased benefits which are difficult to compute, but never the less, include reduced motor maintenance, less wear and tear on equipment and lower starting peak KVA.

Besides the rapid payback of the investment and profits accruing thereafter, a constant cold water temperature will facilitate production planning and control with greater ease and accuracy.

Colder climates where icing is a serious problem causing damage to the tower, VFD control can be automatically sequenced to reverse the tower fans simultaneously (and gently) eliminating monitoring by watch engineers. This retention of the heat periodically inside the cooling tower is one of the best methods of preventing icing in a cooling tower.

Energy savings is tremendous since it is estimated that along the Gulf Coast the potential annual energy savings in cooling tower operations alone is 326,700,000 KWHR per year or 8.16 million barrels of crude oil. This is calculated upon a 30% savings in 500 towers of 300 horsepower at 40 KWHR per barrel of crude.[6] □

Sources Cited

1. Al Kilgore, "Power Electronics and Drive Systems" — Westinghouse Electric.
2. Jim Jerzycke, "What You Wanted to Know About VF Drives; McGraw-Edison Service Co.
3. U.S. Army, Navy, Air Force Engineering Weather Data.
4. Chemical Plant Name & Location on Request.
5. Robert Burger "Cooling Tower Technology" Textbook, 1979.
6. R.W. Corey and J.C. Campbell, "Minimize Energy Consumption With Variable Speed Fan Control," CTI Meeting, 1 January, 1981.

HYDRAULIC COOLING TOWER FAN DRIVES

The prime function of the water cooling tower is to reduce the temperature of circulating water in the cooling systems of modern processes which generate excess heat. When designed and operated properly, water cooling systems allow the primary process to function at an optimal level.

One conventional method of cooling the water is to move air through a cooling tower causing the exchange of heat between the hot water and the colder air. In the past, the cooling air was moved by a rotating fan turned by a fan deck mounted speed-reducing gearbox connected to an electric motor by a shaft and flexible couplings.

In 1986, H.E.M. developed the ***HYDRO-AIR*** cooling tower fan drive which eliminates the shaft, coupling and gearbox. With this arrangement, the electric motor operates at ground level and drives a hydraulic pump which drives a hydraulic motor that drives a fan inside the cooling tower . (See illustrations 1 & 2)

Illustration 1

A - **Electric Motor** (Drives hydraulic pump)
B - **Hydraulic Pump** (axial piston, variable displacement, pressure compensated - pushes hydraulic oil under pressure to the hydraulic motor)
F - **Oil Filter** (cleans hydraulic oil)
C - **Reservoir** (L-typr design to provide flooded suction for pump)
D - **Heat Exchanger** (cools hydraulic oil)
E - **Gauges** (monitors oil pressure - for easy system diagnosis)

Illustration 2

G - **Hydraulic Motor** (low speed, high torque, bi-directional, radial piston - turns fan drive shaft)
H - **Upper Support Bearing** (piloted flange type, generously sized to support fan)
I - **Motor Mount** (designed to fit existing structural members and hole patterns)
J - **Drive Shaft** (turns fan - inside housing- not visable)

The *HYDRO-AIR* system is *SAFE*. Because the electric motor is located on the ground and the gearbox and drive shaft are eliminated, there is no possibility of a dangerous fan wreck or electrically caused fire on the tower. If unforseen problems do occur, the HYDRO-AIR system shuts down automatically before any damage occurs.

The *HYDRO-AIR* system is *Easy to Maintain*. The primary system requirements are clean, cool hydraulic oil and a shot of grease every other week. (This 1/2 oz. shot of grease is the only recommended scheduled maintenance which is done from the outside of the fan stack with a grease gun...*while the unit is running!*)

The *HYDRO-AIR* primary components are *guaranteed for one year*. Similar components have lasted ten years while showing little wear. Actual life of the components depends largely on proper maintenance.

The ***HYDRO-AIR*** system is far more energy efficient than the gearbox and shaft system. (Comparisons available). All cooling tower electric motors and fans are engineered to produce 100% CFM of peak air on demand. However, by definition, "Design Wet Bulb" is needed only 20% of the time. Therefore, 80% of the time a conventional single speed constant RPM fan drive system is overproducing cooling air and wasting energy.

The ***HYDRO-AIR*** system allows the process operator to "Fine Tune" the cooling system to the process requirements. This increase in operating flexibility is due to the inherent variable speeds which the operator can now program into the ***HYDRO-AIR*** fan drives and by automatically adjusting individual fan speeds on multiple cell towers.

The *HYDRO-AIR* system is *easy to operate*. H.E.M. provides complete instructions.

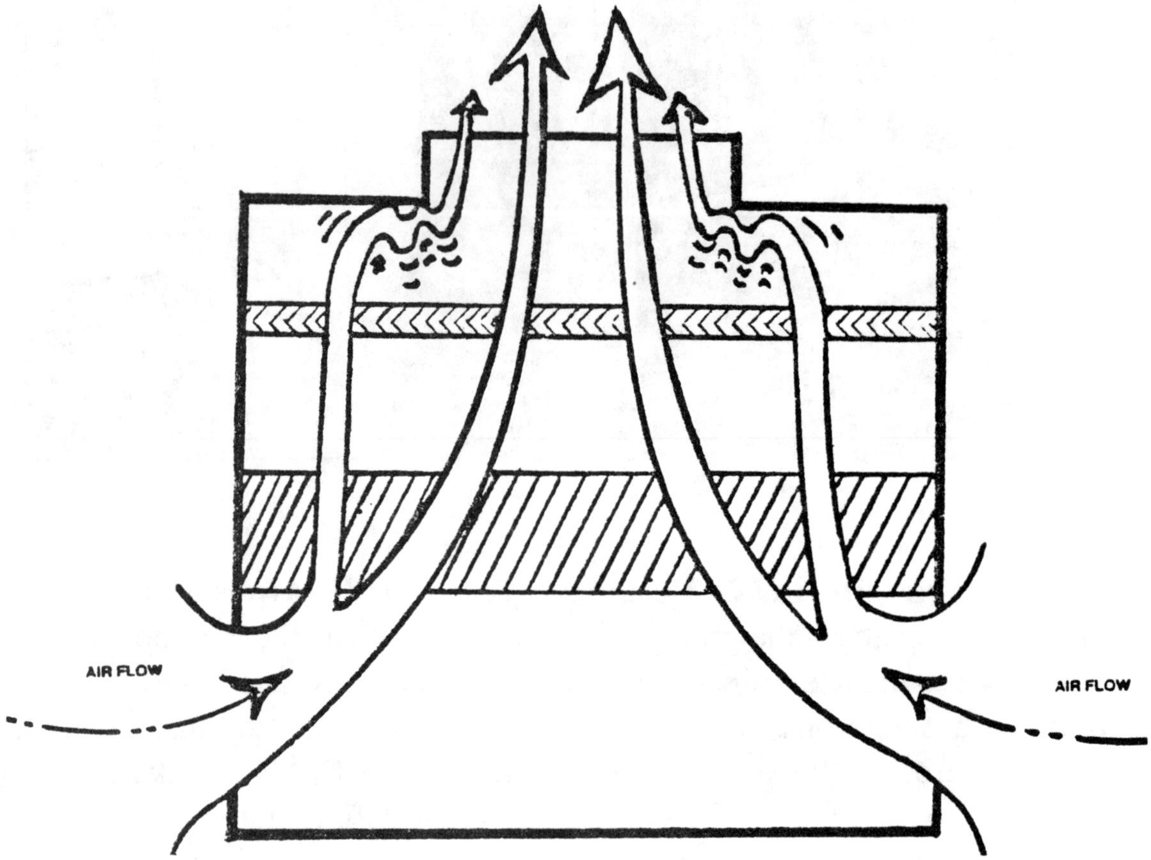

Loss of performance in a cooling tower can be attributed to many factors. Of prime consideration is the kinetic energy loss as the air must maneuver around the right angle turn from the plenum through the fan cylinder.

This is a typical right angle turn required by a conventional fan cylinder which creates resistance, increased static pressure and decreased air velocity. Assume that the fan diameter in this illustration is 14′.

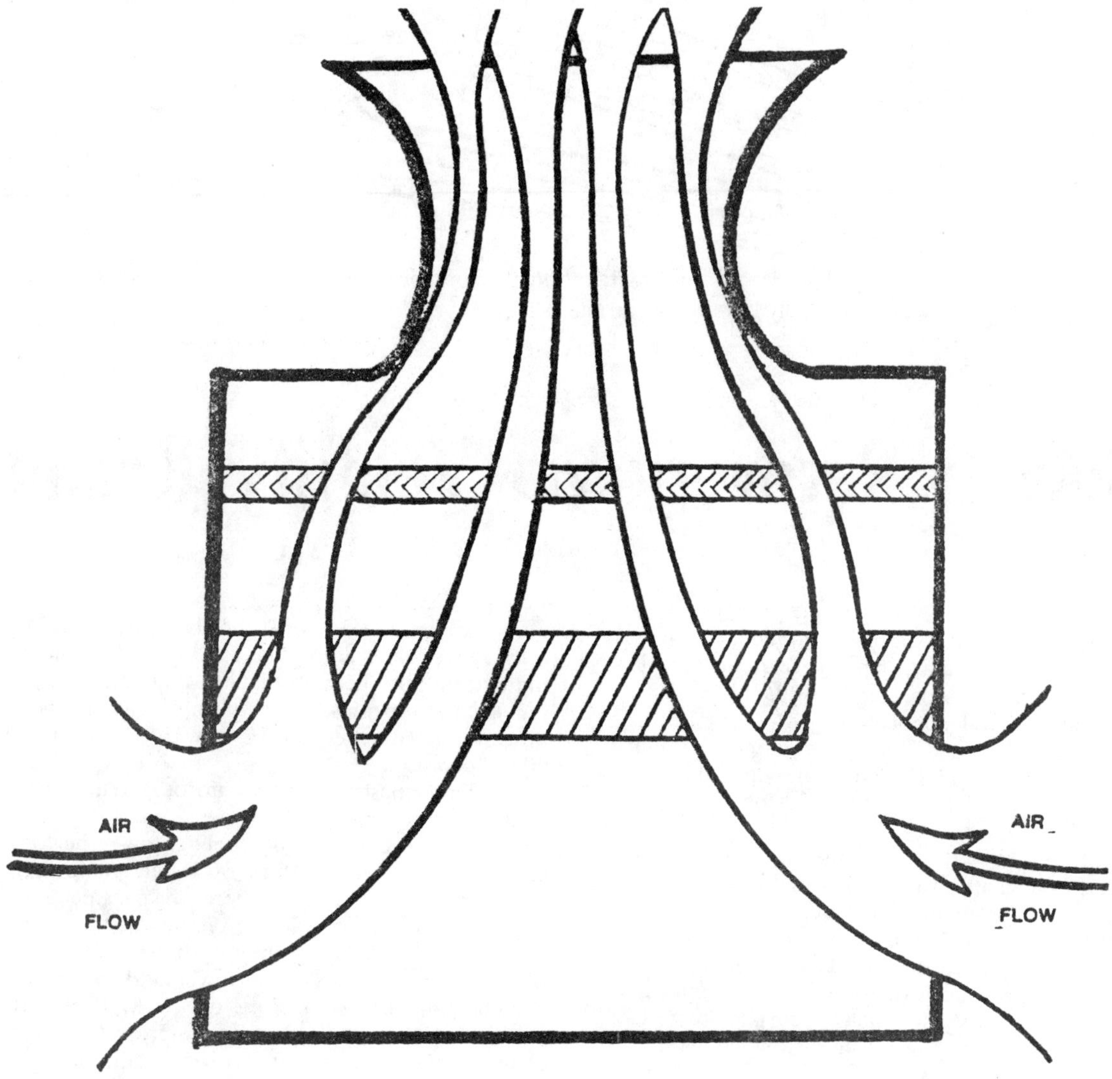

The installation of an eased inlet velocity recovery fan cylinder improves and smooths the air flow exiting through the tower.

The same 14′ diameter as illustrated on the previous page is at the narrow throat of the venturi.

The effect of the venturi reduces the exit pressure and can create upwards to 7% additional air movement at no increase in motor horse power.

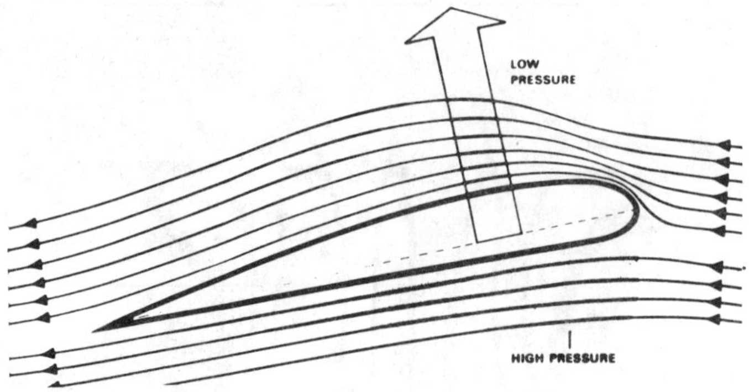

Fig. 1 — Moving air creates low pressure over wing, high pressure under airfoil.

Kilowatt Recovery via Velocity Regain

In today's rapidly changing space age climate, and with new discoveries and practices toppling seemingly solid ideas of the past. Bernoulli's theorem postulated in 1783, still proves its validity and application.

In his studies of fluids and gases in motion, Bernoulli stated that the faster a fluid or gas is moving, the less pressure it exerts on anything with which it comes in contact.

The truth of this is observed in the ordinary act of taking a shower. When the water spray is turned on the shower curtain moves towards the water stream. Why? Before the water is activated, the air (a fluid) is at rest. The water movement increases the velocity of this air thereby reducing the pressure inside the tub. The internal pressure being less, draws the shower curtain towards the area of lower pressure, since fluids flow "downhill" from areas of high to low pressure.

Bernoulli's theorem also explains how an airplane flies. As the propellor or jet engine moves the plane forward, the air flowing over the wing is divided by the leading edge and the particles of air at the leading edge so split, must rejoin at the trailing edge of the wing at the same time. Therefore the lift is not a result of the upward pushing of the air against the bottom of the wing. This is only partially correct; it overlooks the force acting on the upper side of the wing where the bulk of lift is created. In reality lift is actually produced by the differences in pressure of the air above and below the wing which changes its speed of motion. The design of the wing is such that the top curvature, or camber, *(Fig. 1)* is a longer distance than the flat bottom or (chord). Since the air or fluid must move faster across the top to meet its mate traversing the bottom, the theorem dictates that the increased velocity reduces the top pressure, resulting in lifting the wing (and the aircraft attached to it) upwards toward the area of lower pressure.

A good question here would be, "What does this have to do with cooling towers?"

The three elements of cooling tower performance are gallons per minute (GPM) of water, cubic feet per minute (CFM) of air, and temperature parameters. Changing any of these will change the characteristics of the tower.

This article will concern itself with the air generation thereof and behavior in the tower and tower fan cylinder.

The total energy picture of the system must be considered whether it is process water, refrigeration, or air conditioning. Under one set of conditions more energy used by the fan to generate better air is permissible since considerably colder water on, for example, steam driven equipment will net in a lower kilowatt cost to operate the system and less steam consumption.

Colder water means lower head pressures which translated into less power needed to do the job in the engine room. Better air management through the fan cylinder also results in lower horsepower requirements which will be less kilowatt hour consumption thereby saving dollars.

We obtain more effective air in a number of ways:

1. Seal all holes and openings in the fan deck and cylinders to prevent air shortcircuiting and not going through the fill *(Fig. 2 and Fig. 3)*.

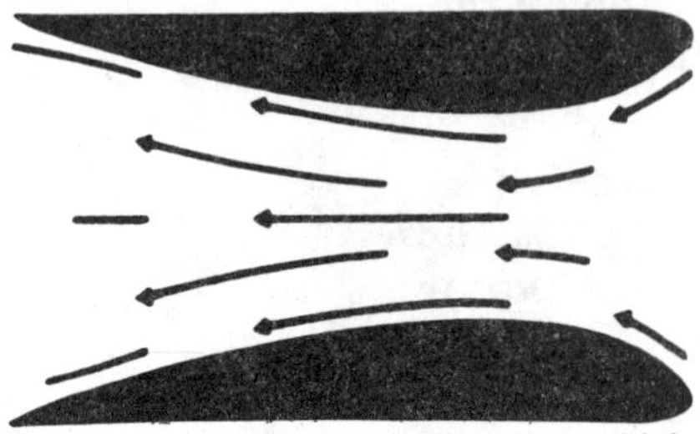

Fig. 6 — Two airfoils form Venturi tube, which is a cross section of velocity regain fan cylinder.

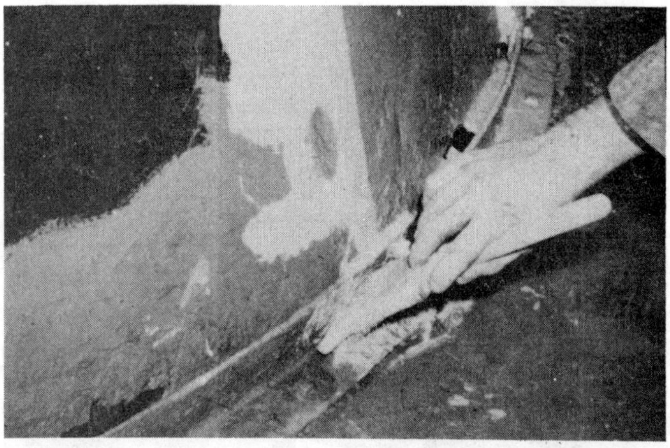

Fig. 2 — Holes in fan stack and basin short circuit air through tower, lowering performance.

Fig. 3 — Fiberglass reinforced with epoxy seals air leaks.

Fig. 4 — Increase air movement by pitching fan blades to maximum amperage.

Fig. 5 — Velocity regain fan cylinder produces more air at less motor horsepower.

2. Pitch the fan blades up to the maximum permissible allowable of the motor plate amperage. *(Fig. 4.)*
3. Utilize velocity fan cylinder *(Fig. 5.)*

It can be seen that if two cross sections of an airfoil *(Fig. 6)* were placed together, a venturi tube would be formed with an area of low pressure generated in the constriction. The region between the cambers would have an increasingly lower pressure depending upon the velocity of air passing through produced by the fan. This utilization of Bernoulli's principle was made in 1822, by Venturi, whose name was attached to the practical application of the theoretical law.

Venturi also laid the ground work for his "tube" to be used as a measuring device by inserting the flaired section in a moving stream and calibrating the pressure differential.

In actuality, this is how the Velocity Regain Venturi Fan Stack operates by seemingly "getting something for nothing."

The following calculations of fan laws illustrate this point:

Given: 192″ diameter 6 blade fan at 239 RPM
50 horsepower motor.
Actual air = 360,000 ACFM
Static pressure = .40″ H_2O (SP)
Velocity pressure = .275″ H_2O (VP)
Actual HP reqd. = 45.32
Total Pressure Actual (corrected) = .695″ H_2O (TPA)

These values are supplied by the fan manufacturer and obtained from his fan performance curves.

Fan Efficiency Total $= \dfrac{(\text{TPA})\ (\text{ACFM})}{(\text{K})\ (\text{HP act})}$

$= \dfrac{(.695)\ (360{,}000)}{(6356)\ (45.32)}$

$= 0.8685$ or 87%

Assume: .80 Velocity Regain with 14′ high Venturi Stack*
(VP) (VR) = .275 x .80 = .22″ H_2O $\left(\dfrac{1}{1.03\ \text{corr}}\right)$
= .60″ H_2O TPA

**From Manufacturers Performance Curves*

Fan Efficiency Total $= \dfrac{(.60)\ (360{,}000)}{6356\ (\text{HP act})} = 87\%$

HP act = 39.10 HP with Velocity Regain

HP Savings $= \dfrac{39.10}{45.32} = 6.22 = 14\%$

Fig. 7 — Elements of VR stack, illustrating existing stack discharge 10′ below building wall.

Fig. 8 — Prevailing wind can no longer cause hot air discharge to recirculate into tower air inlet louvers.

These calculations indicate that a 14% reduction of power and power charges can be effected.

The other side of the coin is that at full power usage, a large percentage of the 14% can be translated into colder water thus either reducing the power consumption of the steam or refrigeration equipment, or will permit more water to be put through the tower.

In many cases, additional load can be put on the tower thereby saving considerable costs of erecting more tower capacity which requires capitalization dollars against VR stack installation which can be charged against the tax deductible maintenance and repair budget.

The Venturi velocity recovery fan cylinders also serve another purpose. *Fig. 7* indicates the elements of the velocity recovery cylinders being installed on a cooling tower. It can be seen that the height of the old wood stacks came to 5′ below the side of an adjoining building. *Fig. 8* is another picture of the same tower showing a DC9 taking off on runway 31 at LaGuardia Airport. This indicates the prevailing winds from the west which blow across the cooling tower and take the hot exhaust air against the side of the building and force it down and into the air intake louvers, re-circulating this hot moist air. Performance was very badly affected by this hot air re-circulation and is now corrected by the velocity regain cylinders.

In other words, besides saving money in horsepower utilization this particular installation gained at least another 15% inefficiency by not sending pre-heated and moistened air through the fill.

On a properly managed installation, it is prudent to investigate the necessity of spending money for increased efficiency. In these days of shortages of all kinds, including power, the engineer with the long-range outlook will always come out ahead.

FIG. 1 — Motors, being the prime mover, should be checked for lubrication, security, and operation.

FIG. 2 — Variable pitch fan blades angled to maximum degree for efficient utilization of motor amperage. Inset: Amprobe on motor leads.

FIG. 3 — When air short circuits, traveling through holes in deck or fan cylinder and not the fill, repairs should be made immediately.

FIG. 4 — Examples of large velocity recovery venturi fan stacks save money through increased efficiency.

How Much Are You Paying For AIR?

■ Air volume and velocity are the work factors in cooling towers. It therefore behooves the operator to investigate the efficiency of generating the air with mechanical equipment, static pressure caused by internal and external sources, and means of improving passage of air.

Cost of generating the air is reflected in power consumption of electric motors. These should be periodically checked for bearing wear, security of connections, proper function of starters, circuit breakers, or fuses, and security of attaching hardware **(Fig. 1)**.

Next item to check is efficient performance of the motor driving the fans to generate air. This is accomplished by pitching the fan blades to the maximum angle allowed by the motor plate amperage. The first step is to check amperage on the motor leads while the fan is under load. If there are amps to spare, pitch the airfoils up accordingly and recheck amperage. By trial and error consumption can be adjusted to within a few percentage points of the plate rating, since the motors have a 10% safety factor built in **(Fig. 2)**.

If existing fan stacks or roof deck have holes in them, these should be patched to prevent air short circuiting. This is an obvious waste of energy, because air drawn from the top of the tower passes out the cylinders without cooling water **(Fig. 3)**.

The largest area of improvement involves installation of velocity regain stacks. Venturi restriction increases airflow velocity through this area by reducing pressure, permitting more air to be drawn through the tower for greater cooling **(Fig. 4)**.

The total energy picture must be considered whether the system is used for process water or refrigeration. Additional energy used by a fan to generate better air is justifiable, because colder water will result in a lower kilowatt cost to operate the system at less steam consumption. For example, on overhead catalytic condensors in chemical plants, 2°F colder water to the "cat cracker" could result in significant additional daily production. Efficient air management through the stacks results in lower horsepower requirements and head pressures, decreased kilowatt consumption, and cost reduction.

Another area in which tall velocity regain stacks may be used is in recirculation prevention. Some locations may have obstructions that cause discharge air to be shunted downward to air intake louvers. This hot, moist air, recirculated with its high wet bulb temperature, can reduce efficiency of a tower as much as 30%. Installation of the V.R. fan stack will pay for itself in a short time through colder water production and increased efficiency.

Field investigation by a qualified engineer could return dividends in daily efficiencies that will return installation costs through reduced operating expense.

CHAPTER 8

STRUCTURAL ANALYSIS

STRUCTURAL ANALYSIS AND DESIGN OF WOOD FRAMED COOLING TOWERS

There is no intent in this seminar to produce structural design engineers. We would like you to be aware, however, that after the design conditions of performance have been outlined, the structural details involve a lot of knowledge and calculations, and the concern of the manufacturer is that his product must be consistant with the required strength together with economies of the competitive market.

After identifying the various loads, numerical values must be placed on them and the resultant forces determined.

Once the design requirements have been defined, the engineer must determine and classify the types of loads imposed on the structure, primarily in terms of load duration. This procedure is necessary, since the allowable stresses on a member will vary with the type and duration of loads imposed. The load classifications to be considered are:

1) Dead Load
2) Operational Loads
3) Live Loads
4) Wind Loads
5) Seismic Loads

Transverse Axis:	Plane parallel to the end walls.
Longitudinal Axis:	Parallel to louver face.
Columns:	Vertical load bearing members.
Girts or Joists:	Diagonal braces providing lateral stability.
Beams:	Horizontal load bearing members.

DEAD LOAD:

Weight of all components, structural and operational. Weight of framing material, fill, supports, drift eliminators, mechanical equipment, fan stacks, water distribution piping. Dead loads are considered long term loads.

OPERATIONAL LOAD:

Weight of water in piping, water in hot water basins, and water suspended in the fill plus ice loads, axial fan thrust, and fan out of balance forces.

LIVE LOAD:

Weight of repair or service equipment temporarily on structure as well as personnel loads on decking or access walkways.

WIND LOAD:

Determined by height gradient and are computed based upon effective wind areas of the exposed shapes and spacing of wind bracing elements. The wind force at any given point is simply the wind pressure at that level times the effective wind area.

SEISMIC LOAD:

The variation is due to the local seismic history in terms of zone factors.

After recognizing all of the distributed forces, the engineer must then combine the various classifications of loads to determine the maximum loading condition on each member joint. The combinations of loads will vary from project to project but the basic calculations are the same.

The columns (Figure 1) must resist either axial forces or axial forces in combination with bending forces as in the case of louver columns where the direct incidence of wind acts as a uniform load. Horizontal members, such as girts and joists (Figure 2), must resist bending forces due to a uniform or concentrated load and may have to resist simultaneous axial forces originating from lateral loads. Diagonal basin members (Figure 3), generally must resist only axial forces originating from lateral loads and provide lateral stability for the entire structure.

Once the magnatude, direction, and types of loads have been identified and quantified, the engineers must size members and design joints to resist these loads.

ANALYSIS - COMBINED LOADS

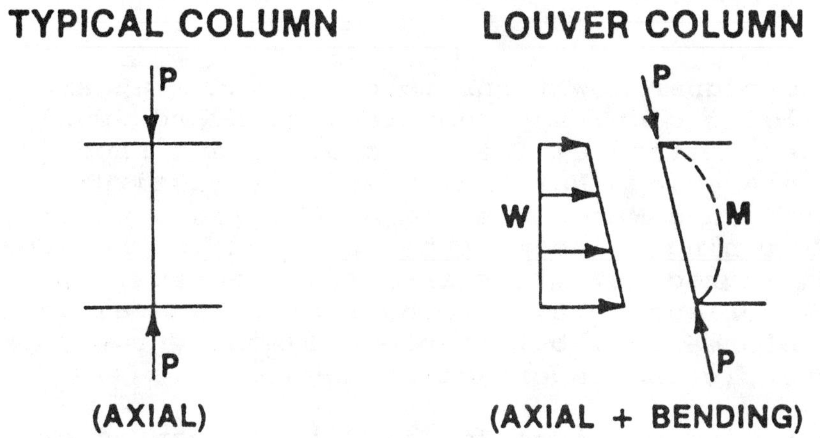

FIGURE 1

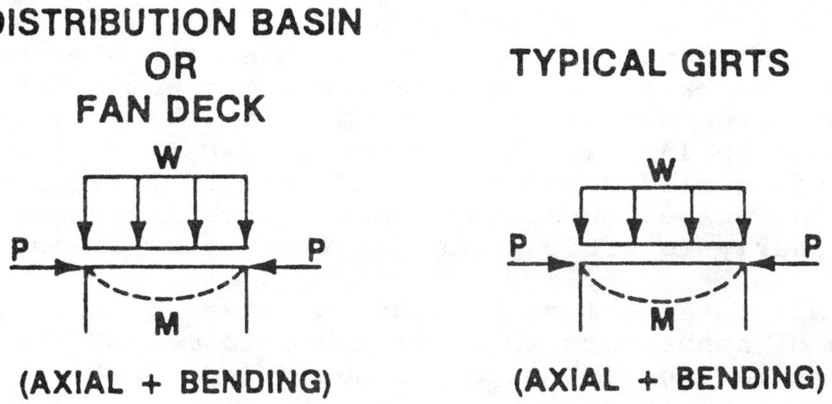

FIGURE 2

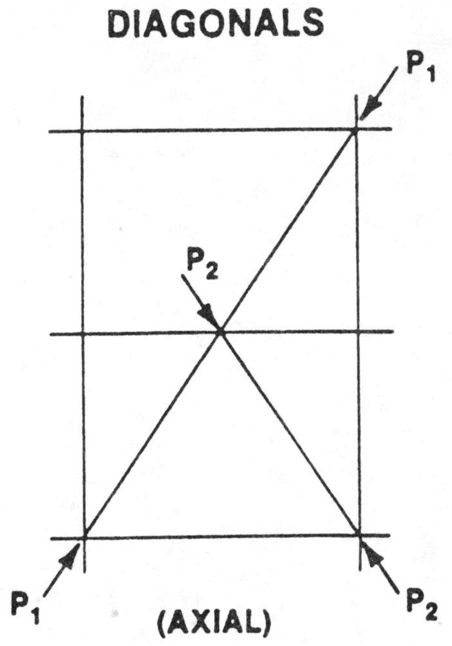

FIGURE 3

The structural designer now turns to codes and standards such as the uniform building code, supplementary standards published by The Cooling Tower Institute, The National Forest Products Association, The American National Standards Institute, American Society of Testing and Materials, to name a few. Other reference books and local zoning codes must be referred to; however, there are two missing ingredients that are still lacking. These are knowledge and experience. All of the textbooks available have the necessary information, but to properly apply them, the experience factor is the important catalyst.

In designing for wind pressure loading, the accepted criteria is 100 miles per hour maximum velocity. The wind pressures shown on the left side of Figure 4, combined with the concept of effective wind areas, are used to develop the specific wind loads imposed, while the transverse line of bracing at each framing level. In this case, the forces resulting from a 100 mile wind criteria range from 1,416 pounds of the level of the hot water basin to 625 pounds at the basin of the structure. It must then be determined what the applied loads produce in the form of transverse loads on the individual structural members. For example, the top most diagonal is subject to a force of 2,160 pounds, while midway down the tower, the member force increases to 6,079 pounds and finally to 8,329 pounds at the bottom of the diagonal.

Additional calculations must be made to determine the type, size, and strength of connectors to be utilized to resist these loads generated at a maximum 100 mile per hour wind.

After analysis of all of the forces involved and considering the many interdependent parameters, the structural engineer must then produce a resulting design that is both structurally sound yet economically practical.

APPLIED WIND LOAD
NO DIAGONAL FORCES

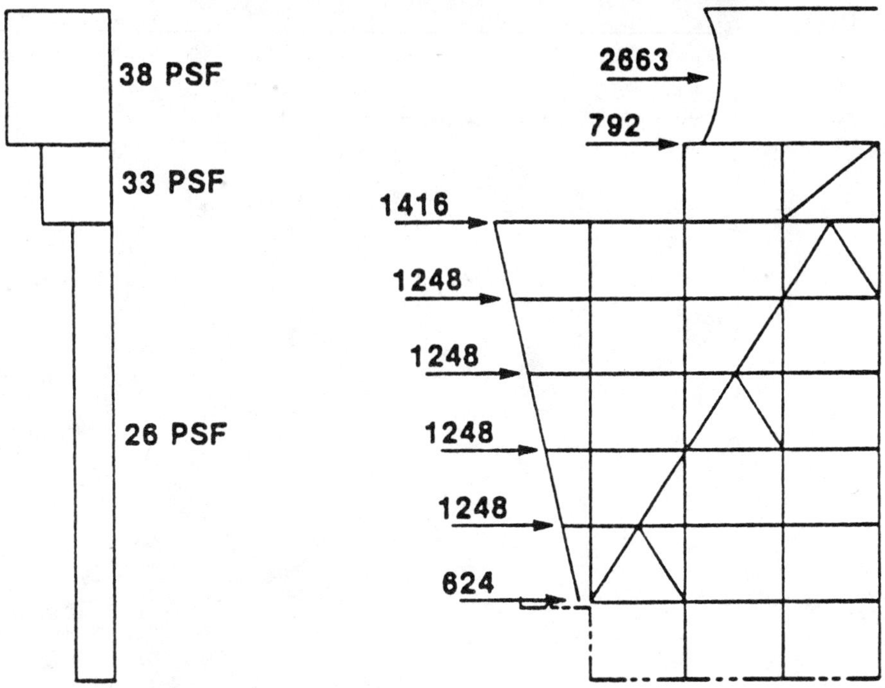

DISTRIBUTED WIND FORCES
WITH DIAGONAL FORCES

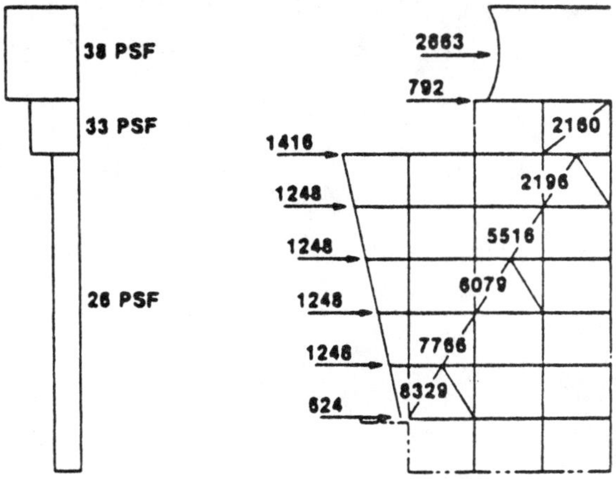

FIGURE 4

CHAPTER 9

NATURAL DRAFT COOLING TOWERS

NATURAL DRAFT COOLING TOWERS

Natural draft cooling towers have been a familiar sight in Europe for many years and only recently been appearing on the American scene. At first, these towers were constructed of wood, then of wood over steel. Modern natural draft towers are generally constructed of reinforced concrete. The shape of the tower has also changed over the years; early towers were cylindrical, and later a design using two truncated cones was used. The current shape is hyperbolic, giving the tower superior strength and an ability to match the natural flow of air through the tower shell. Less material is required for construction because less total tower volume is involved in the modern designs.

Natural draft towers do not use fans to move air; the flow of air through the tower is created by the density difference between atmospheric air and air inside the tower, which is warmed as the water is cooled. Hence, the tower is little more than a chimney which draws cool air in at the base and exhausts warm air at the top. Two basic airflow schemes are used in relation to water; crossflow and counterflow, and these are illustrated in Fig. 1. Counterflow towers provide the more efficient means for heat transfer, although the better uniformity of air and water distribution achieved in crossflow towers may in practice offset the disadvantage. In the crossflow design, airflow is normal to the water movement and usually more fill is required in the counterflow tower; the crossflow tower, however, has a lower pressure drop. A further engineering and cost factor to be considered, however, is the pumping head which in counterflow towers is nominally less than in crossflow. In view of the large volumes of water (100,000 to over 350,000 GPM) this pumping head power cost must be evaluated seriously. Fig. 2.

FIG. 2. Water volumes 100,000 to 500,000 GPM surges through the tower like the "tide coming in".

CAPITAL COST CONSIDERATIONS

For a given cooling duty, mechanical draft towers can be built with fewer and less expensive materials than natural draft towers. The mechanical draft design obviously eliminates the need for a "chimney section", which represents 30% to 50% of the cost of a natural draft tower. Although fans are required for mechanical draft towers, this cost; including wiring, controls, and operating expenses is more than offset by the very large initial erection cost. While the mechanical draft tower consumes power, on a twenty year life span, the power input to fabricate the materials of the natural draft tower are at a higher level.

In addition to basic construction costs, land costs must also be considered for specific sites. Mechanical draft towers must usually be removed from the immediate plant facility (and from each other) to minimize the effective recirculation and local fogging. The higher exit point of warm air from a natural draft tower obviates most of these types of problems. Over a certain size, it is customary practice to build the cooling duty into two or more separate tower structures for mechanical draft systems. The required separation of the towers can thus cause the mechanical draft tower system to occupy four to five times as much land as the same sized natural draft tower systems; there is a corresponding additional expense for land. Another added cost is, of course, represented by the extra piping that connects the various cooling tower structures with the process plant.

Because of these considerations, there is a capacity level at which the mechanical draft system becomes more expensive than a natural draft system. The type of cost relationship shown in Fig. 3. will, therefore, be

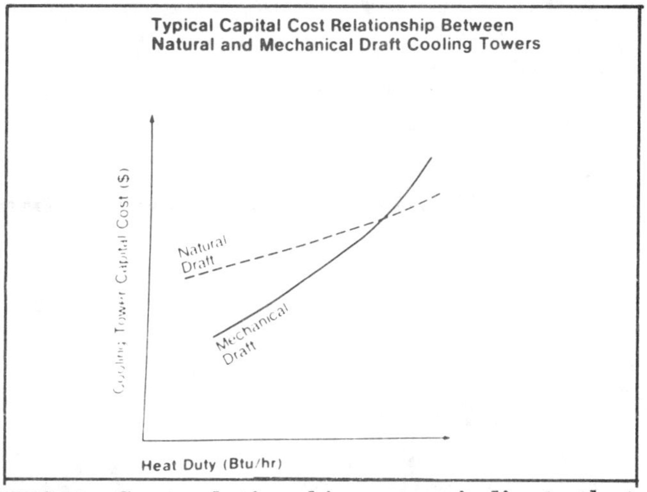

FIG. 3. Cost relationship curves indicate that in the area of 5,000,000 BTU per hour heat removal, a decision between natural draft and mechanical draft towers should be made.

Natural Draft Cooling Tower

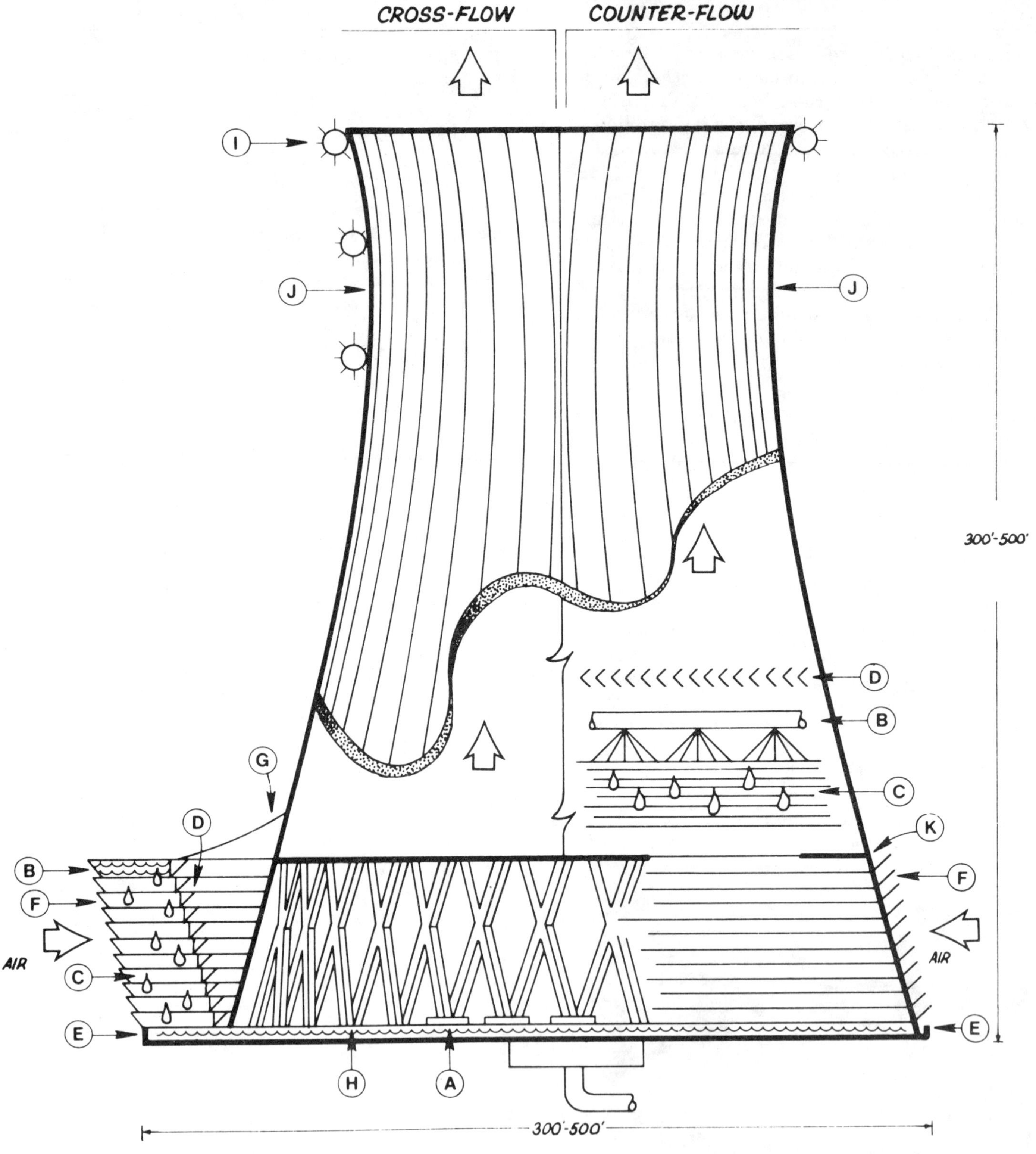

A-PEDESTAL
B-WATER DISTRIBUTION
C-FILL PACKING
D-DRIFT ELIMINATORS
E-COLD WATER BASIN
F-AIR INLET LOUVERS
G-CANOPY
H-CROSS STRUTS
I-OBSTRUCTION LIGHTS
J-THROAT DIAMETER
K-SOFFIT

FIG. 1. Natural draft cooling towers can be either counterflow or crossflow.

observed. As cooling duty increases, the cost differential between the two decreases until the natural draft tower becomes the more economical choice in the long run. Where this point occurs depends on Design Conditions at the individual site and will depend on specific land costs and site preparation requirements. From the limited data assembled, it appears that the cross over point occurs at a cooling duty that will be in excess of 5,000,000 BTU's per hour.

GEOGRAPHICAL FACTORS

Local climatic conditions play a large part in the selection of wet cooling towers. Because Petrochemical and other similar industries must design for summer conditions and their loads remain relatively constant throughout the year, mechanical draft towers are generally a better choice. Mechanical draft towers are usually recommended where a low approach is required (from 10 deg F to 20 deg F).

The natural draft tower has its best application in the power industry where winter loads might exceed summer loads; the total heat load is very large, and pay out period is long. In general, natural draft towers are recommended where operating conditions consist of a low wet bulb temperature and high inlet and outlet water temperatures (i.e., broad cooling range and cooling approach), or where there is a heavy winter load. Low relative humidities in Southern California, Arizona, Utah and Nevada preclude its use in such areas because of excessive water loss from the recirculating system by evaporation. Winter load conditions and moderate relative humidities in the Northwest might favor natural draft towers based on process considerations. In the Gulf Coast area, natural draft towers may be unacceptable because of high air temperatures in the summer.

Natural draft towers are best suited to the Northern part of the Eastern U.S. where relative humidities are generally high. Fig. 4 shows the location of large natural draft cooling towers for electric utility use, either in service or under construction. The greatest number of towers is in West Virginia, Pennsylvania, Ohio and Kentucky, with a few others found in Georgia, Virginia, Tennessee, Arkansas, California and Oregon.

The state of the art of cooling tower technology is constantly changing and improving. This is particularly pertinent to the natural draft cooling tower experience. Many plants are totally dependent on their cooling tower for the success of their operation—it is an important and essential piece of equipment in the power generating industry.

This very expensive equipment is purchased to do a particular job called Design Conditions. Design Conditions consist of five variable factors namely, gallonage of water to be cooled, temperature on the tower, temperature off the tower, and ambient wet bulb which is made up of relative humidity and dry bulb temperature.

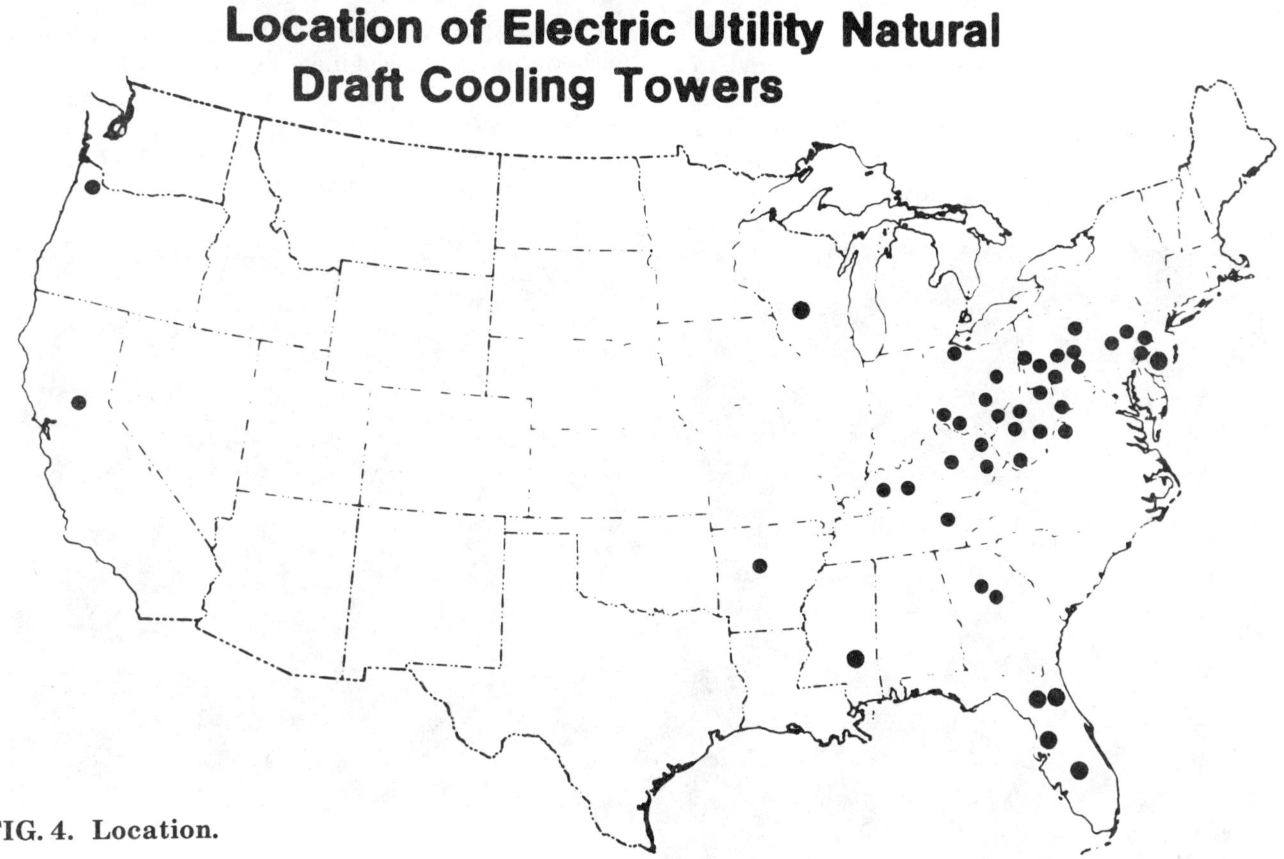

FIG. 4. Location.

COOLING TOWER INSTITUTE

The Institute is a non-profit self governing technical association. It is dedicated to improvement in technology, design, performance and maintenance of cooling towers. In addition, water and air pollution have always been and will continue to be a prime concern to CTI and members.

With so many dollars at stake it seems incredible that the users of this very expensive structure do not call upon the Cooling Tower Institute to conduct impartial unbiased tests of the multimillion dollar natural draft tower to verify its pcrformance.

The cost of a natural draft tower can be as high as $15,000,000 each and mechanical draft installed prices, depending upon the number involved, can range upwards to over $1,000,000. CTI Code calls for uniform methods and instrumentation for all conditions and measurements involved in determining tower cooling performance. Fig. 5.

CTI has a contract aggreement with Midwest Research Institute (MRI) of Kansas City, Missouri, to provide technical services for CTI. These services include testing of towers as requested by manufacturers and users. The MRI representative makes the calculations required to analyze the thermal performance of the cooling tower and reports the test results only to the manufacturer and the tower purchaser/user. This testing service is arranged for through the CTI Houston Office, 9030 North Freeway #216, Zip 77037 (713)445-9431. attention Dorothy Garrison.

Purchasers of industrial cooling towers are urged to specify a CTI CODE TOWER to insure purchase of 100% thermal performance and the highest quality product. They may be purchased anywhere in the world and afford the best assurance of obtaining a quality product.

The cost of this testing, depending upon the size of the tower should be in the area of $2,000.00. Regardless of who contracts for the test, to maintain the 100% unbiased objectivity, the owner receives a copy of the test results directly from the Midwest Research Institute.

All users, manufacturers, and suppliers to the cooling tower industry are urged to apply for membership in the Institute. In order to protect the Owner's investment so that he will receive what he is paying for, the specification should state, "This is to be a CTI CODE TOWER". Immediately nine CTI standards become part of the specifications and contract with the Manufacturer. These standards, if applicable are:

- STD-103 Redwood Lumber Specifications
- ATC-105 Acceptance Test Code
- STD-111 Gear Speed Reducers
- STD-114 Douglas Fir Lumber Specifications
- STD-115 Southern Pine Lumber Specifications
- STD-118 Inquiry and Bid Form
- STD-119 Timber Fastener Specifications
- STD-127 Asbestos Cement Materials for Application on Industrial Water Cooling Towers
- STD-201 Certification Standard for Commercial Water Cooling Towers

CTI Standard ATC-105 describes the CTI CODE TOWER Acceptance Test Code which has become a standard in the industry. This test code is used to determine the water-cooling capability of cooling towers.

By associating with the CTI, Operating Management and Maintenance Engineering personnel will have available to them, and will participate in the latest cooling tower state of the art. This capability will further enhance the professionalism of the Engineers and financially benefit his or her Employer.

CHAPTER 10

COOLING WATER CHEMISTRY

The necessity of understanding the chemistry of cooling tower water treatment reflects itself primarily in economic considerations and also in compliance with Environmental Protective Agency (EPA) requirements. Heat transfer between fluids and metals, in addition to corrosion and deterioration of the equipment are of significant financial import in water reuse. Also the EPA imposed thermal pollution controls together with limitations on discharge of water treatment chemicals in the blowdown to cooling tower operations multiplies the cooling tower responsibility.

Designers of cooling systems and their components, cooling tower operators, regulatory officials, and others who should or may wish to become familiar with this increasingly complex and important subject should be conversant with the fundamentals of cooling tower water technology.

This section is neither capable of, nor intended to make any one an expert, but it should provide a better understanding of the causes of waterside problems and some guidance in the selection of procedures for their prevention or correction.

The subject can be examined by looking at cooling tower operations from the viewpoint of the circulating water, seeking answers in turn to each of the three questions:

1. What happens to the water used in cooling tower operation?
2. How does cooling water quality affects the efficiency, reliability, and life of cooling systems?
3. What can be done to minimize undesirable water caused effects on cooling system operation and life?

It is to be noted that chemical treatment of the cooling water for corrosion control purposes is not to be intended for the preservation of the cooling tower's ferrous metallic components or structures. These parts, since they are alternately wet and dry, and sometimes, while the tower is not operating for periods of times, are subject to atmospheric and chemical attack. Protective coatings and maintenance, as discussed in other sections, are essential for the conservation of the cooling tower.

1. COOLING WATER CONTAMINANTS.

Water is probably our nearest approach to a universal solvent, dissolving a little of almost everything with which it comes into contact—oxygen and other gases when it is falling as rain or as droplets in a cooling tower, and almost every material over and through which it flows while passing to the river, reservoir, or well from which it is drawn for cooling tower use. The dissolved and suspended "impurities" in the water, rather than the water itself, are the sources of problems in water-using systems.

Water Analysis Units—These impurities are present in quite small amounts. The concentrations of dissolved and suspended materials in typical fresh waters are so low that the water chemist expresses them as parts per million (ppm) or the metric equivalent, milligrams per liter (mg/l), rather than percent, in order to avoid the use of decimal points with a string of zeros before each significant number. To put these units in perspective, one percent equals 10,000 ppm or mg/l in fresh waters. For more highly concentrated waters, such as brackish or sea waters, ppm must be multiplied by the density in g/ml for conversion to mg/l. Trace constituents are often expressed as parts per billion (ppb) or the metric equivalent, micrograms per liter (ug/l).

The terms used to report water analyses are not completely standardized, so it is important to include an identifying term along with the numerical value as, for example, "mg/l as $CaCO_3$" or "ppm as Cl." Given the identifications, it can quickly be determined that three calcium analyses reported as 100 ppm as $CaCO_3$, 40 ppm as Ca, and 56 ppm as CaO are in fact identical even though the numbers are quite different.

The concentrations of dissolved and suspended materials in cooling tower makeup waters may range from almost zero in systems which utilize steam condensate as makeup to the 30,000–50,000 ppm range in cooling towers which use sea water as makeup. In between, there is a broad concentration range for the ground and surface waters that are used as makeup for most cooling towers and, in the upper portion of this range, for the re-used waters or waste water treatment plant effluents that are being used increasingly for cooling tower makeup.

Water analyses always include the pH, a frequently misunderstood dimensionless unit which expresses the hydrogen ion concentration or, in plain language, the strength of acid or alkali. Note that this is *strength* and not the *concentration*. The latter is reported as ppm of acidity or alkalinity. pH is expressed on a logarithmic scale which ranges from zero at the strong acid end to 14 at the strong alkali end, with 7 the nominal neutral point. However, because much of water chemistry involves the equilibria among the various dissolved forms of carbon dioxide (carbonic acid, bicarbonate ion and carbonate ion), the water chemist usually considers pH 4.3, below which there is no more bicarbonate ion, as the boundary between acidity and alkalinity. The term "free mineral acidity" is used by water chemists to differentiate the acidity caused by such strong acids as sulfuric and hydrochloric from that represented by the carbonic acid present between pH 4.3 and 8.2.

Water Composition—A very complete analysis of almost any cooling tower makeup water would disclose the presence of scores of constituents including dissolved solids and gases, colloidal solids, and suspended solids. For example, the U.S. Geological Survey reports the concentrations of 44 constituents in the water supplies of the 100 largest American cities in its Water Supply Paper 1812. Fortunately, only a limited number of these are really significant insofar as effects on the operation and life of most cooling systems are concerned and the usual water analyses report only these.

Among the more important *dissolved solids* are total hardness, calcium, alkalinity, and sulfate. These are the most frequent sources of scale formation in cooling systems. Silica is a less common source of scale. Phosphate may cause scale formation when the makeup is a treated wastewater. Other common deposits may form by precipitation from the makeup water of iron or manganese as their hydrated oxides. Chlorides intensify corrosion under various circumstances, but they also serve as a useful water treatment control tool for measuring the degree to which dissolved solids have concentrated in evaporative cooling systems because they are too soluble to form deposits and treatment processes do not usually affect them significantly.

$$\frac{\text{ppm Chloride in tower water}}{\text{ppm Chloride in makeup water}} = \text{Cycles of Concentration}$$

The three *dissolved gases* most commonly analyzed for in water supplies (carbon dioxide, oxygen and hydrogen sulfide) all promote corrosion, as does sulfur dioxide which is sometimes absorbed from polluted air during cooling tower operation.

The usual water analyses also include color and turbidity, which are rough measures of *suspended and colloidal materials*. Such materials may include fine silt or clay particles, oil, colored natural organic materials from the decomposition of leaves and other forest materials, living organisms such as bacteria and spores, and other materials. Each of these can contribute to deposit formation or otherwise adversely affect the operation or life of a cooling system.

Changes in Water Composition—The potential water-caused effects in any cooling system cannot be predicted solely on the basis of the *makeup* water analysis. Cooling tower operation changes the water composition quite radically, and it is the *circulating* water composition that really affects the system's operation and life.

Water used in cooling tower operation is exposed or subjected to three, and sometimes four, processes which alone or together change its physical properties or chemical composition. These processes are:

—Temperature change,
—Evaporation,
—Air Contact, and
—Product Contamination

Even should the makeup water itself be harmless to the cooling system, these processes will almost universally produce changes that will adversely affect the system operation or life unless preventive or corrective steps are taken.

As the heat transfer medium in a cooling tower, the water enters the tower at 10°F or more above the *temperature* at which it leaves. A water that is benign at the lower temperature may become significantly corrosive or deposit-forming at the higher temperature.

Heat rejection in the tower is brought about by *evaporation* of a small percentage of the cooling water. Water alone evaporates, leaving behind the solids that were dissolved and suspended in it, thus building up their concentrations in the remaining circulating water. This process can and commonly does convert soft, corrosive makeup waters to scale-forming circulating waters.

Cooling towers are designed to break up the circulating water into fine droplets so as to maximize evaporation. This also happens to insure that the circulating water will be saturated with oxygen at all times, thereby increasing its corrosivity, and that it will efficiently scrub other gases and particulate matter from the air passing through the tower. An industrial environment may include such gases as sulfur dioxide, ammonia, and hydrogen sulfide and such solids as silt, fly ash, salts, bacteria, and spores of other living organisms. Depending upon which *materials are scrubbed from the air*, the cooling water may significantly increase either its corrosivity, or its deposit-forming or fouling potential.

The contrasting effects produced in cooling tower water by the normal evaporative cooling process and by scrubbing of air containing excessive amounts of stack gases may be seen from the analyses in Table I of water samples from an industrial plant in south central Pennsylvania. No water treatment program was in use for either cooling tower during the first several weeks of initial plant operation and both cooling systems suffered water-caused shutdowns. Water in the office building tower showed a typical pattern of increased pH, together with much lower

Table I. Effects of Evaporation and of Flue Gas on Water in Untreated Cooling Towers

(Figures in parentheses are ratios of constituent in tower water to that in the makeup water.)

Constituent	mg/l as	Make-up Water	Cooling Towers	
			Office	Plant
pH	—	7.5	8.3	4.3
Alkalinity	$CaCO_3$	136	182 (1.3)	0 (0)
Free Acidity	$CaCO_3$	0	0	2
Chloride	Cl	11	205 (18.6)	931 (84.6)
Sulfate	SO_4	47	950 (20.2)	3500 (74.5)
Hardness	$CaCO_3$	184	761 (4.1)	860 (4.6)
Calcium	$CaCO_3$	124	716 (5.8)	780 (6.3)
Iron	Fe	0	0.1	250

concentrations of alkalinity, hardness, and calcium than would be expected on the basis of the cycles of concentration calculated from the ratio of chloride in the tower water to that in the makeup water. This indicated heavy calcium carbonate precipitation which was confirmed by the heavy scale found in the refrigerant condenser.

A very different picture was presented by the water in the large plant cooling tower located on a roof less than 100 feet away and, unfortunately, only 10 feet from the stubby plant stack. Scrubbing of the stack gases resulted in the absorption of enough sulfur oxides to exceed the neutralizing capacity of the alkalinity in the makeup water. The excess sulfuric acid which formed drastically reduced the pH. The corrosivity of the circulating water is obvious from the excessively high iron content. The very high sulfate concentration also caused heavy calcium sulfate deposits.

Finally, changes in the cooling water composition are sometimes caused by *leakage* from higher pressure process streams in heat exchangers. Here, too, the water may become more corrosive or develop a greater fouling tendency, depending upon the nature of contaminants.

2. EFFECT OF WATER QUALITY.

A cooling tower is part of a carefully engineered heat exchange system. Any film or deposit which forms on the waterside heat exchange surfaces reduces the heat exchange efficiency of the system. System reliability can be sharply reduced by maintenance shutdowns required for removal of waterside deposits, replacement of circulation pump shaft seals abraded by suspended particulates in the water, or repairs required by waterside corrosion failures. Corrosion of system components shortens the system life, as well as reducing operational reliability.

The cooling tower operator sees these water-caused effects as symptoms—cooling capacity becomes inadequate—cooling water flow decreases—leaks develop. However, these same symptoms may also develop from causes unrelated to the cooling water. Insulating deposits may form on the *process* side of heat exchangers. Thin plastic sheeting has been known to enter a cooling tower and cut off water flow by blanketing heat exchanger tube inlets. Leaking pump glands may reflect poor mechanical maintenance. So, the first step in correcting these situations must be a diagnosis of the causes of the symptoms.

Reduction of Heat Transfer—But even when it is established that a problem is water-caused, it must be further defined before an effective remedy can be selected because the same symptom may develop from different causes, each of which requires different corrective steps. Thus, waterside reduction of heat transfer rate can be caused by buildup of any material on heat transfer surfaces. The fouling deposit may result from sedimentation or adhesion of suspended material; formation of calcium carbonate, calcium sulfate, or silica scale; or accumulation of bacterial slime. All of these waterside foulants reduce heat transfer, but different methods are needed to prevent the deposition of each.

Sedimentation of suspended solid materials, such as the fine sand frequently present in well waters, silt and organic matter present in many surface waters, corrosion products or mill scale particles generated in the system itself, precipitates formed in water treatment processes, or particles of silt or fly ash scrubbed from the air in open circulating systems, usually produces a deposit along the lower parts of horizontal pipes or tubes. But it is also possible for some solids to accumulate all around the tube circumference, as in the cases of corrosion products remaining on the surfaces at which they are formed, oily leakage from process streams adhere in to all exposed surfaces, and biological growths clinging to any available surface.

A surprisingly small amount of biological material, such as *bacterial slime,* can sharply reduce heat transfer by entrapping and holding immobile a very much larger volume of water, as well as by entrapping particles of sediment.

Scale is the most widely-known water-caused trouble. Indeed, the term is so familiar that it is often applied incorrectly to *all* solid accumulations in water systems. The correct generic terms are deposit or fouling, of which scale is but a single type; others being sediment, corrosion products, slime, etc. A true scale forms by crystallization of a dissolved salt when its concentration exceeds its solubility. In contrast to other water-caused deposits, a scale will re-dissolve without chemical change if the water composition is so altered that the water becomes unsaturated with respect to the precipitated compound. Most scale-forming materials of significance in water technology

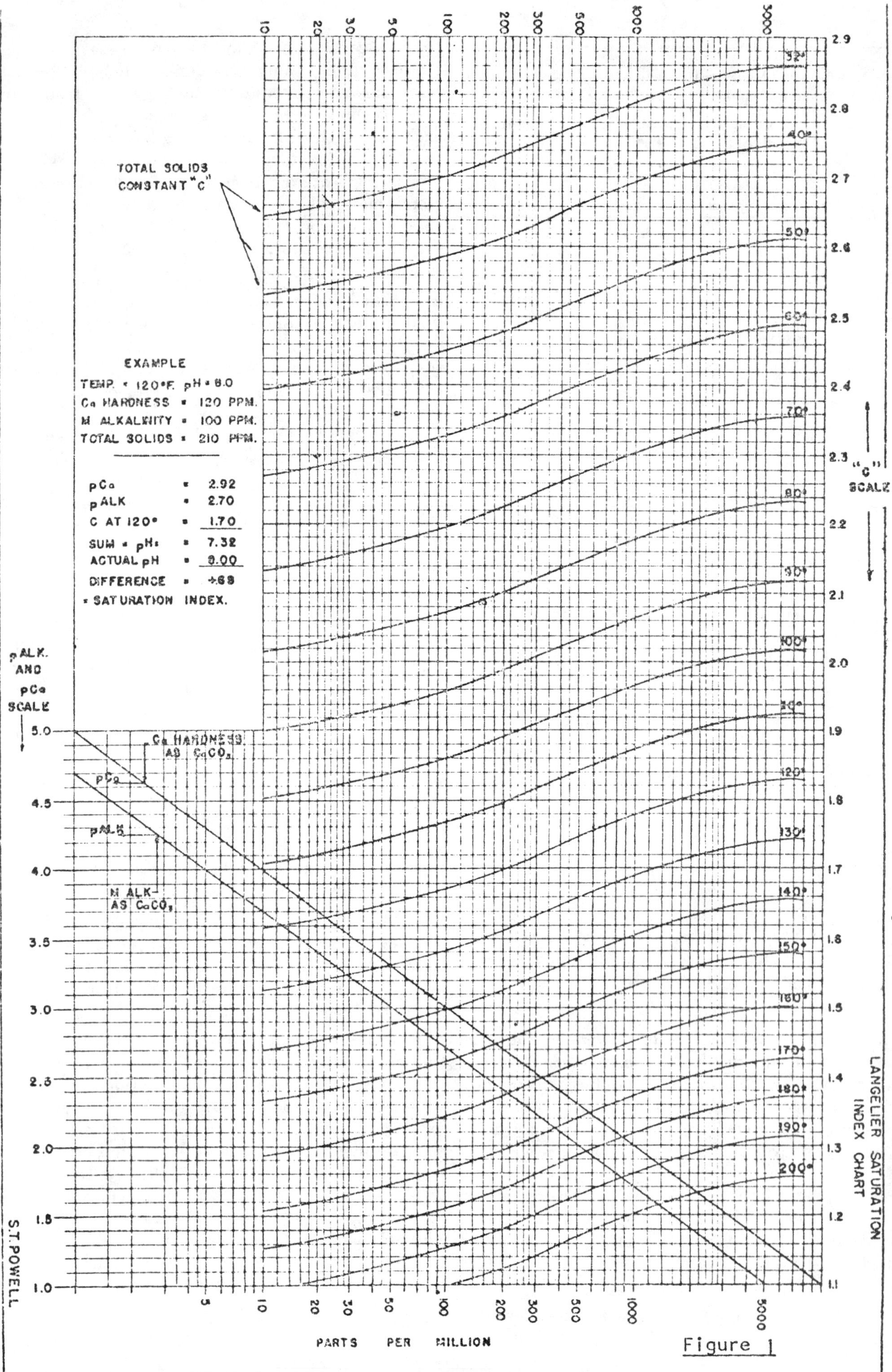

FIG. 1. Langelier Index which assists in determining the pH in water.

have the unusual property of becoming less soluble as the water temperature increases so that scale tends to form on the heat exchange surfaces—the very locations which should be kept deposit-free for efficient operation.

By far the most commonly formed scale consists of calcium carbonate, the chemical of which limestone is formed and for that reason often referred to as lime scale. Its constituents—calcium and alkalinity—are present in all natural waters. For any given water composition and temperature the tendency to form calcium carbonate scale can be calculated from the temperature, pH, and the concentrations of calcium, alkalinity, and total dissolved solids by the use of formulas or diagrams based upon the work of Langelier. Figure 1, for example, may be used to determine pHs: the pH at which a water of this composition would be in equilibrium with calcium carbonate at the indicated temperature. The Langelier Index itself is defined as pH minus pHs.

$$\text{Langelier Index} = \text{pH} - \text{pHs}$$

If this index is positive, the water tends to form calcium carbonate scale; if negative, the water tends to dissolve calcium carbonate. But note that the Langelier Index indicates only the *tendency* to precipitate calcium carbonate. It does *not* tell whether or not a scale will actually form and, if it does, how fast it will form. Thus, various treatment chemicals and some natural organic materials present in many surface waters act as sludge dispersants and prevent scale from forming even when the Langelier Index indicates that there is a strong scale-forming tendency. Also, between about +0.5 and −0.5 or possibly a somewhat broader range, the Langelier Index is unreliable for predicting whether or not any scale will actually form even in the absence of scale inhibitors.

Furthermore, and contrary to common belief, the Langelier Index does *not* tell us whether or not a water is corrosive. It is generally *assumed*, with varied degrees of correctness, that the more negative the Langelier Index, the more corrosive the water.

Ryznar calculated the same pHs and combined it with the pH in a different fashion to give a Stability Index which he then correlated with experience data to give a somewhat better picture of what may be expected of a given water under the conditions assumed for estimating the pHs.

$$\text{Ryznar Stability Index} = 2\,\text{pHs} - \text{pH}$$

As shown in Figure 2, a Ryznar Stability Index below 6.5 generally indicates a scale-forming tendency which becomes more pronounced as this Index decreases, and as this Index rises above 6.5, the waters are usually increasingly corrosive to steel.

Non-carbonate scales are deposited in water systems when the solubility products of calcium sulfate, calcium phosphate, or silica are exceeded. In cooling tower practice, calcium sulfate is the most frequent scale-former after calcium carbonate, particularly in systems in which sulfuric acid is used to adjust the pH for scale control. Calcium sulfate is, however, far more soluble than calcium carbonate: about 1200 ppm as compared with about 50 ppm for calcium carbonate. Nevertheless, the maximum cycles of concentration in the cooling tower water is at times limited by the danger of calcium sulfate scale formation. Makeup waters high in silica, particularly those containing upwards of 40 to 50 mg/l SiO_2, are a source of silica scales and in such waters the silica content may limit the cycles of concentration. Calcium phosphate deposits can be a problem in systems supplied with secondary sewage treatment effluent as makeup.

Denman has published a procedure for calculating the calcium sulfate precipitation tendency of a water (Ind. Eng. Chem. *53,* 817–22 (October 1961)). Green & Holmes have published one for calcium phosphate (J. Amer. Water Works Association. 39, 1090–96 (November 1947)), and Larson et al have published one for magnesium hydroxide (ibid. *51,* 1551 (1959)).

Reduced Water Flow—Accumulations of the same sediments, corrosion products, scales, or biological growths which cause reduced heat transfer can also reduce the flow of cooling water at times. In order to restrict flow, however, the deposits are usually much heavier and develop at locations other than the heat transfer surfaces. Thus, a heavy growth of algae in the distribution basin of a cooling tower can block water distribution nozzles, or scale or tubercles of corrosion products may build up in piping. At times, blockages can develop quite rapidly, as when pieces of scale break away from one location and lodge in fittings or valves, flakes of corrosion products clog strainers, sheets of plastic film are drawn into a cooling tower and block off a number of condenser tubes, or careless workmen leave tools in a cooling water line.

Deterioration of Equipment—Chemical, biological, or physical conditions, or combinations of these can each lead to the deterioration or breakdown of cooling system components. Equipment failures may be sudden, as in the bursting of a chiller tube when the chilled water freezes because of mechanical malfunction. More commonly, cooling system components deteriorate gradually as a result of such chemical action as the corrosion of metallic components, microbiological deterioration of cooling tower wood, or physical wear as in the abrasion of pump shafts or mechanical shaft seals by solid particles suspended in the water.

The many forms of corrosion are by far the greatest causes of cooling equipment deterioration and take place to some extent whenever a metal, water, and a

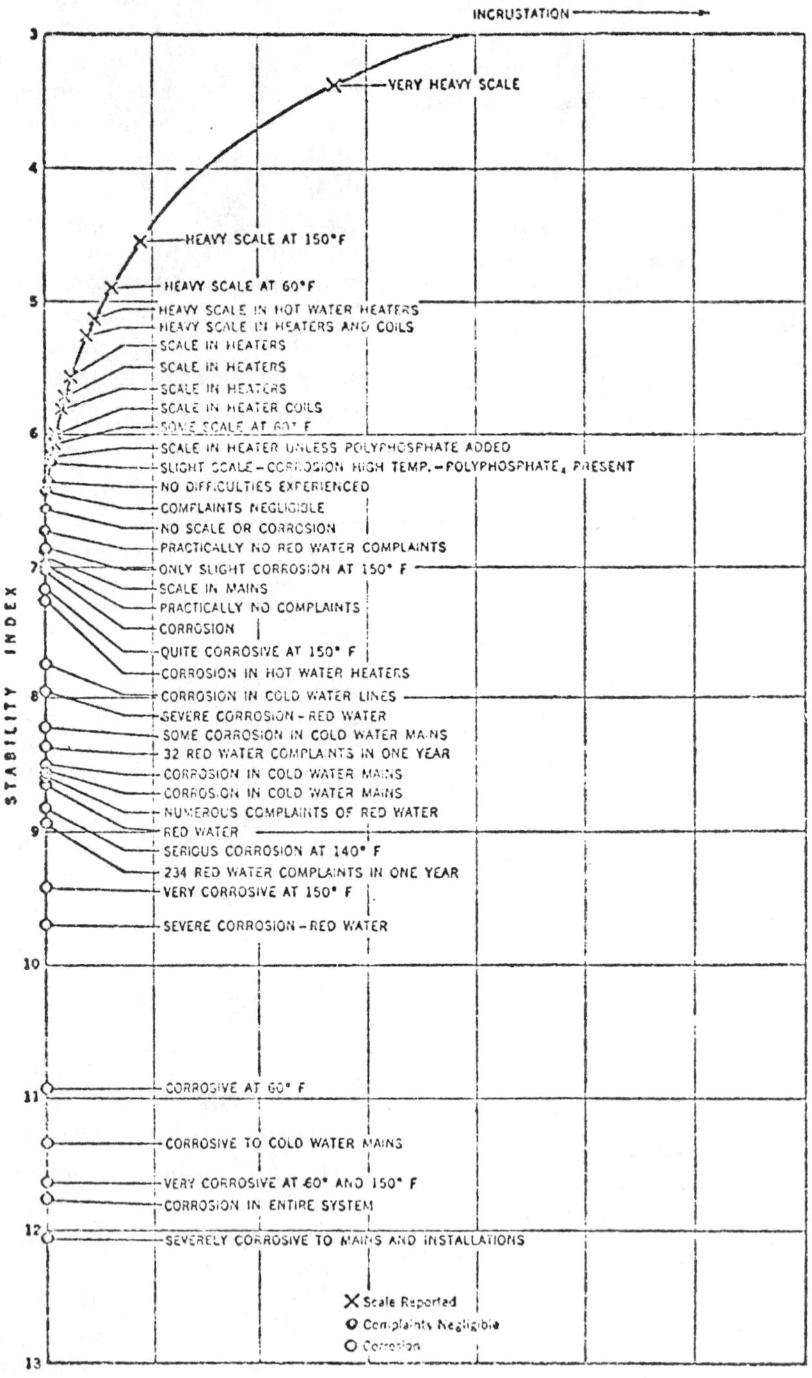

FIG. 2. Ryznar stability index for determining relationship between pH and scale-forming tendencies.

depolarizing agent, such as the oxygen which dissolves in cooling water as it passes through the tower, are brought together. Since corrosion is a chemical process, its rate increases with temperature. Low pH, which can develop by pickup of sulfur dioxide or other acidic gases in the tower or of acidic process fluids in heat exchangers, also accelerates corrosion. Non-chemical factors including among others, water velocity and deposit formation, play a considerable part in localizing corrosion, thereby accelerating its effects.

Because corrosion is electrochemical, any condition that causes one point on a metal surface to differ from another will permit establishment of a localized corrosion cell. The galvanic action caused by joining two different metals in the presence of water is well-known. The relative areas of different metals that are connected in a water system also affect the corrosion damage. A large area of anodic (more corrodible) metal connected to a small area of cathodic (more resistant) metal is commonly acceptable because total corrosion is spread over a large area. Thus, brass valves are commonly and safely used in steel piping systems. On the other hand, a reversal of the relative areas of the two metals can lead to a catastrophic rate of attack, as when an all-brass or copper piping system includes a single steel nipple.

Less well-known, but no less common and damaging, are localized corrosion effects produced by crevices, deposits of any type, local stresses caused by cold working, and many others. Low water velocities permit suspended solids to settle out, and localized corrosion can then take place under the resulting deposits, and this is one of the more common causes of localized corrosion in cooling systems. On the other hand, excessively high velocities can produce impingement attack, which is common and particularly rapid on copper and many alloys.

Although we usually think of corrosion in terms of metals, water can also cause other materials of construction to deteriorate. Cooling tower lumber can be delignified chemically by excessive chlorine concentrations or weakened by the action of fungi. Concrete is attacked by soft waters and acidic waters.

3. MINIMIZATION PROCEDURES.

The prevention or correction of the many undesirable effects that can be caused by cooling tower water requires consideration and action during the earliest stages of design and throughout the construction of the cooling system, as well as during its operation. Selection of suitably resistant materials and tube sizing to avoid impingement attack are two decisions which must obviously be made during design. Some corrective measures help to avoid more than one problem, as when biocidal treatments prevent formation of slime deposits which both reduce heat transfer and cause localized corrosion. Some problems require more than one corrective step. Thus, calcium carbonate scale is prevented by a combination of limiting the concentration of total dissolved solids in the circulating water, reducing the alkalinity by adding acid or reducing the hardness by softening the makeup water, and adding a scale inhibitor or dispersant to the circulating water. Measures needed for the prevention of some problems may aggravate others, as when the reduction of pH to minimize calcium carbonate scale formation tends to increase the corrosion rate. Any proposed water treatment procedure must first be evaluated for its effects on the total operation of the system.

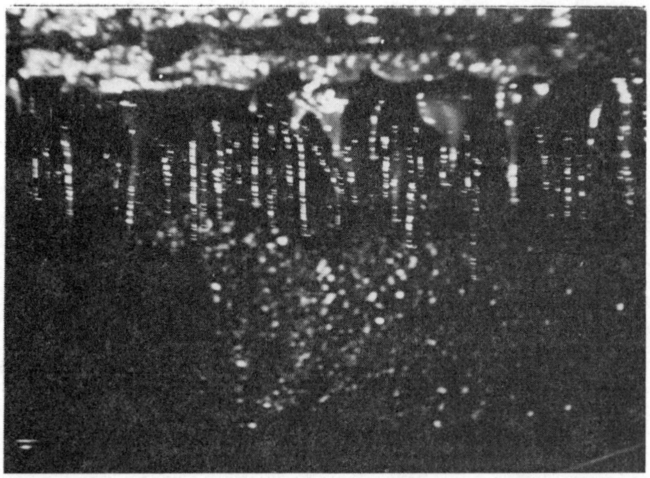

FIG. 3. Bacterial slime hanging from the bottom of wood fill in a cooling tower.

Biological Growths—Slime and algae in cooling systems are frequently spoken of and handled as though they were a single word and single problem. That this is not so may be demonstrated by the ease with which algae, which require sunlight for their growth, can be prevented in many cases by installing an opaque cover on the cooling tower distribution basin, although this action has no effect upon the formation of bacterial slimes in the system.

With the exception of such algae control steps as supplying opaque covers for the distribution basins of smaller cooling towers at some locations, and siting or providing louvered sun shields to minimize exposure to sunlight, there is little that can be done during design and construction in order to prevent microbiological growths during operation. Shut-downs caused by such growths can be minimized by providing convenient access, hose connections, and drains for physical removal of algae and slime accumulations from cooling towers and back-flush connections for their removal from heat exchangers, particularly for systems in which there is a good probability of biological problems developing, such as those in which sewage treatment plant effluent will be used as makeup. Minimizing the development of algae and bacterial slimes is important because they reduce heat transfer, decrease cooling water flow, localize corrosion, and serve as a mortar for permitting rapid buildup of deposits consisting of an agglomerate of sediments, corrosion products, and scale. The aim in cooling tower treatment is microbiological control to avoid significant slime deposits or algae growth, not the almost impossible goal of maintaining a completely sterile circulating water.

Treatment for microbiological control is becoming increasingly difficult as a result of pollution control regulations and water conservation measures. The latter lead to operation at higher cycles of concentration, which increases the concentration of nutrients in the circulating water, particularly in some industrial environments, such as ammonia plants. At the same time, pollution control regulations limit the toxicity to aquatic life of any blowdown discharged to a river, lake, or the sea. Here we must face up to the paradoxical requirements for microbiocides. They must be toxic to the undesirable organisms in the cooling water, but at the same time be non-toxic to the organisms—large and small—that live in the water to which the blowdown is discharged. In some cases the receiving water may be a biological wastewater treatment plant, in which case the microbiocide must kill undesirable bacteria in the tower but not be harmful to useful bacteria in the wastewater treatment process. The importance of considering the total cooling water system—water source, water and air conditions in the recirculation loop, and the system receiving the blowdown—when selecting a water treatment cannot be pointed up better than by these situations.

The many treatment chemicals available for microbiological control in cooling towers include chlorine and other compounds which yield available chlorine, a variety of quaternary ammonium compounds, organo-tin compounds, chlorophenates, organic thiocyanates and thiocarbamates, and many others. Some of these compounds have broad spectrum effectiveness and others are specific for a more limited range of organisms.

Chlorine and other chlorine-yielding compounds are both broadly effective and widely used. Their future use is likely to be restricted by EPA limitations because very low concentrations of chlorine have been shown to be toxic to some forms of aquatic life in waters receiving cooling tower blowdown and because of recent concerns about the formation of traces of chlorinated organics that may make the water less desirable for potable use. The utility of other commonly-used biocides faces restrictions through pollution control regulations (chlorophenates) or changes in treatment technology for scale and corrosion control (thiocyanates).

The relatively high cost of microbiocides, the large and constantly changing variety of organisms present in any cooling tower and the ability of colonies of organisms to gradually accommodate to any given biocide influence biocide treatment and control techniques. In contrast to the preferred practice of proportionately feeding deposit and corrosion control chemicals, biocides are usually slug fed at intervals and it is common practice to alternate the use of two different biocides or to change biocides from time to time.

It would be virtually impossible to maintain a sterile cooling tower water at all times. And although some organisms are much more troublesome than others, the time and cost for regularly monitoring the population of particular species present in a cooling water is so high, that this is rarely done for treatment control. Instead, biological control is most frequently maintained by periodic measurement of the overall bacterial population of the water through a standard plate count, sometimes called a total plate count. Most cooling systems will be free of problems caused by organisms if the standard plate count is below 500,000/ml. Most require prompt microbiocide treatment when this count rises above 1,000,000/ml.

One additional control procedure is frequent, often daily, inspection of the tower distribution basin for signs of increased algae growth.

Finally, whenever flagrant microbiological growths develop, chemical or mechanical cleaning must be included along with microbiocide treatment in any effective program for promptly re-establishing microbiological control.

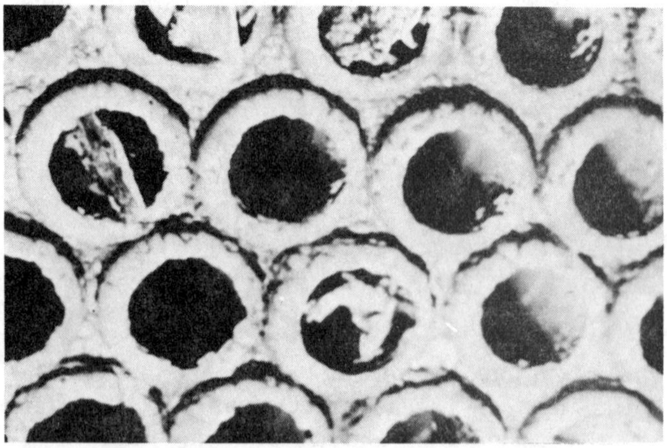

FIG. 4. Tube sheet of heat exchanger indicating substantial deposit of hardness scale on tube walls, ends, and sheet.

Deposits and Scale—Fouling control requires several different approaches. As pointed out earlier, control of bacterial slimes contributes to minimizing one type of deposit formation; so do corrosion control measures, since iron rust and other corrosion products are common constituents of many cooling system deposits. But the major causes of deposits are the settling out of suspended particles from the cooling water and the growth of calcium carbonate or other scales.

Deposit formation resulting from the *sedimentation* of suspended particulate matter in the circulating water can be minimized by preventing its entry into the system, by removing it from the circulating water, or by keeping it dispersed throughout the water.

Settleable solids may enter the tower as sediment in the makeup water. Preventing this entry by coagulation and filtration of makeup water drawn from turbid rivers is practiced in many larger cooling tower installations. Sand in well water makeup supplies may be removed by passage of the water through centrifugal separators known as hydrocyclones.

It is also common for sediment to enter a cooling tower by scrubbing of particulate matter from the air passing through a cooling tower. Although filtration of the influent air is usually uneconomic because of the cost of installing and cleaning air filters, and of the additional power needed for blowers, in some specialized situations intake air filters have been needed. The clogging of strainers by matting of fur fibers scrubbed from the air required such installations in several comfort air conditioning cooling towers in the fur district of midtown New York City.

Removal of sediment from the circulating cooling water by filtration of the total flow is both excessively costly and unnecessary. When deposits tend to form by settling of suspended matter, the usual technique is to keep suspended matter in the cooling tower water down to an acceptable level by passing a portion of the circulating water through filters or hydrocyclones. A so-called slip stream or side stream, usually between 1% and 5% of the total flow, is passed through a sand filter or other granular particle filter, a tubular strainer, or a hydrocyclone. This technique of putting a ceiling, usually less than 100 ppm, on the amount of suspended solids in the circulating water in order to minimize sedimentary deposits is analogous to the blowing down of a small percentage of the circulating water in order to limit the dissolved solids concentration so as to prevent scale formation.

Sediment may be dispersed in the circulating water so that it remains suspended and is removed from the system as part of the blowdown. The effectiveness of dispersion depends upon the characteristics of the sediment and of the system. In an attempt to minimize settling, heat exchanger designs usually provide for cooling water flow rates in excess of 1 ft/sec in tubes and 3 ft/sec in shells. In recent years polyelectrolytes have become available which keep sediment from settling out when fed to the circulating water at concentrations on the order of 1 ppm.

Scale control is simple in theory. Limit the concentration of the potential scale-former to an amount below that at which it will begin to deposit, add to the water a material which inhibits scale formation or growth, or both. Concentration control can be achieved by several different procedures. For example, calcium carbonate scale can be prevented by reducing the Langelier Index of the circulating water; reducing the calcium through softening the makeup water; reducing the alkalinity and pH by feeding acid; or limiting the calcium, alkalinity and dissolved solids of the circulating cooling water through blowdown. The scale control program which was most commonly used in industrial cooling towers until recent years consisted of feeding sulfuric acid both to reduce the alkalinity and to adjust the pH to approximately 6.5, maintaining a blowdown to limit the total dissolved solids, and adding a polyphosphate, lignin or other scale inhibitor to maximize dissolved solids concentration that could be maintained without scale formation.

FIG. 5. Tube sheet of heat exchanger shortly after draining and opening. Good control of water treatment indicates clean surfaces free from significance scale and corrosion products.

Common practice with this treatment program was to design for a circulating water Langelier Index of about +1.0 or a Ryznar Stability Index of about 5.5 to 6.0. This established the maximum cycles of concentration in the circulating water and from this the required blowdown can be calculated from a materials balance on dissolved material in the makeup and blowdown water. The simplified blowdown equation is:

$$B = E/(C - 1)$$

where B is the blowdown rate (or, more precisely, the combined blowdown and drift rates), E is the evaporation rate, and C is the maximum cycles of concentration selected on the basis of the Langelier or Ryznar Index calculation.

This program illustrates the interplay of treatment programs with other problems. The addition of acid, usually sulfuric, permits conservation of both water and treatment chemicals by permitting operation with higher cycles of concentration. But in many waters the fullest extent of such conservation cannot be achieved because the sulfuric acid addition increases the potential for calcium *sulfate* scale formation to the point where this, rather than possible calcium *carbonate* scaling, becomes the factor limiting the cycles of concentration. Also, the pH range desired for preventing calcium carbonate scale tends to increase corrosion rates in the system, thus requiring the use of more effective inhibitors which, in many cases, introduce environmental problems in the disposal of the treated blowdown.

When a high sulfate makeup water or the addition of sulfuric acid makes calcium sulfate scale formation a possibility, it is possible to calculate the maximum cycles of concentration to avoid such scale formation, but this is rarely done. Instead, as a sufficiently accurate maximum, the cycles are limited to a point at which the sum of the calcium and sulfate (both as $CaCO_3$) is below 1500 ppm or at which the product of these is below 400,000–500,000.

Similarly, a calcium phosphate limit can be calculated to provide the maximum cycles of concentration when using sewage treatment plant effluent as makeup.

High silica makeup waters can lead to formation of very difficultly removable silica scale. At present, not enough is known about the mechanism by which these form to select the limiting cycles of concentration with as much assurance as is done for the other common scale-formers. In general, silica scale can usually be avoided by limiting the silica in the circulating water to a maximum of about 150–200 ppm.

In recent years there have been developed several families of more effective crystal growth inhibitors and dispersants; particularly the polyelectrolyte polymers, organic phosphonates, and phosphate esters. By preventing deposition of calcium carbonate or calcium sulfate scales even when the circulating water conditions are such that the solubility product of either compound has been grossly exceeded, these have caused a trend away from the use of acid feeding for scale control to so-called high pH programs in which the circulating water seeks its own pH level. With most makeup waters this is in the 8.0–8.5 range as a result of the equilibrium established between the carbon dioxide in the air and that in the circulating water. Like the polyphosphates which preceded them as scale dispersants, these are usually effective at concentrations of only a few ppm of active material.

Corrosion—Control of corrosion in cooling tower systems begins with materials selection and other design factors. Indeed, with the EPA discharge regulations barring the use of any corrosion inhibitors in future cooling towers in the steam electric generating industry, proper materials selection and other design considerations will be about the only way to effectively control corrosion in power plant cooling tower systems. In other cooling tower systems, the normal economic factors still apply so that treatment for control of extensive runs of mild steel piping or heat exchanger tubing is necessary in most systems.

FIG. 6. Upper and lower portions of copper alloy heat exchanger tube indicating pieces of sediment and corrosion products deposited in the line along the bottom as a result of insufficient velocity.

Until recent years, the chromates were widely used for this purpose. Indeed, we still have no other cooling water corrosion inhibitors that are as effective for the protection of virtually all metals within the pH range encountered in cooling towers. A variety of inhibitor systems have been developed which, with careful treatment control, provide satisfactory inhibition when using extremely small concentrations of chromate. Beginning with polyphosphate-chromate mixtures which gave effective protection in certain pH ranges with about 30–40 ppm chromate, it has been found possible to further reduce the chromate concentration by using mixtures with zinc salts, phosphonates, polyacrylamide copolymers, and mixtures of these. This, together with higher cycles of concentrations and higher pH range resulting from more effective scale inhibitors, has reduced the chromate discharge problem to manageable limits in many cases. In some larger installations, removal of chromate from blowdown before discharge has proved to be economically acceptable. However, analysis of such cases shows that most involve special circumstances so that it is not reasonable to expect these techniques to be widely used, particularly in moderate or smaller-sized cooling systems.

The trend toward higher pH circulating waters mentioned above and pollution control measures are making it more difficult to use zinc in corrosion inhibitors. But the higher pH trend has reduced overall corrosion rates caused by the circulating waters, and the improved materials and techniques for preventing deposits of suspended matter or biological growths has reduced the opportunities for localized corrosion to develop. The use of these improvements permits the attainment of satisfactorily low corrosion rates using corrosion inhibitors which are not quite as effective as the chromates, but which are more acceptable environmentally in the blowdown discharge. Most of these inhibitors are proprietary organic chemicals which may include phosphonates, phosphate esters, silicates or polyacrylates. Generally, these require higher dosage than inorganic inhibitors.

Because of the variety of conditions encountered in most cooling tower systems, it is still necessary to apply other corrosion control techniques for the protection of specific areas within many systems. Among these are the designed control of water velocity to minimize impingement attack, the use of tube-end inserts for minimizing corrosion within the first few inches of the inlet ends of heat exchanger tubes, the use of sacrificial anodes for cathodic protection of water boxes, and the use of protective coatings for steel cold water basins of cooling towers and other portions of some cooling systems.

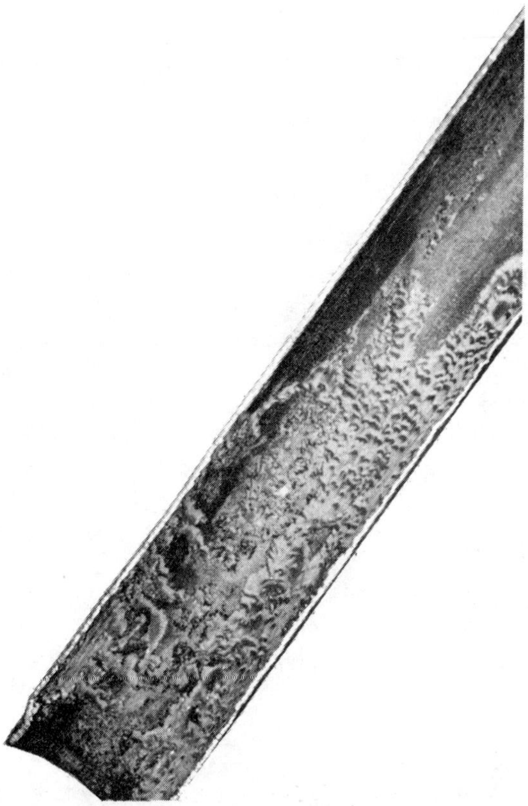

FIG. 7. Inlet end of heat exchanger tube showing severe impingement corrosion resulting from local high turbulence and velocity.

CHAPTER 11

SITING

SITING OF COOLING TOWERS

Fig. 1. Most industrial cooling towers are sited without consideration of optimizing performance. Usually piping requirements dictate the location.

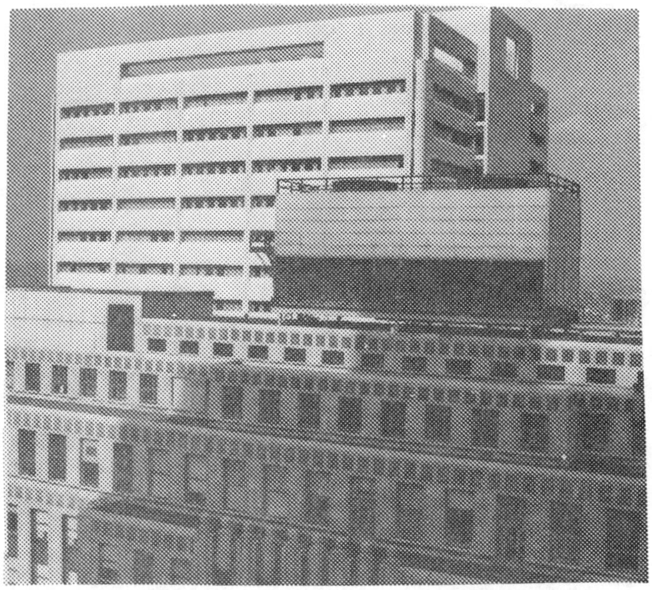

Fig. 2. Refrigeration cooling towers are sometimes located on inaccessible roofs due to plot restructions.

More often than not, a cooling tower is erected on a plot of ground or put on dunnage beams with little or no thought to maximizing it's location or efficiency of operation.

Industrial towers for chemical process or electric generating power plants are usually put in the back, "out of the way", with little thought or consideration except for the ease of piping it. (Fig. 1.)

Refrigeration/air conditioning towers have a greater availability of mis-application when it comes to location. They can be placed:—

1. In a penthouse
2. On a roof. (Fig. 2)
3. On the ground.
4. Underground.
5. In an enclosure (Fig. 3.)

Other options, depending upon aesthetics, would be to make the cooling tower part of the building. This works very nicely from the architectural point of view, but places a strain on the engineering for proper operation.

A cooling tower selection guaranteed to produce Design Conditions is predicated upon a sufficient quantity of air entering the tower and an unobstructed discharge.

Three basic configurations of cooling towers are used and each has it's own advantage together with it's disadvantages.

FORCED DRAFT TOWERS

These types of structures, also called "blow through towers", (Fig. 4) have the air movers located at the bottom of the tower to force the air through it. The characteristic of this design is such that high inlet velocities occur with resultant low velocity exit air. Forced draft towers are lower in height and usually operate at higher horsepower requirements than other configurations. They require unobstructed air inlet and discharge areas.

Improper location of a forced draft tower can create very sub-standard results which are costly and embarassing to specifying engineers. Such problems do not build the owner's or architect's confidence in their consultants.

COUNTERFLOW TOWERS

Industrial counterflow induced draft towers and smaller refrigeration units, because of the fan location on top, achieve a high discharge velocity of hot, moisture laden air from the fan stack which can vary between 1500 to 2000 feet per minute. The discharged air is directed up and away from the intake sections of the tower which tends to provide a more efficient cooling tower design.

There is more flexibility in locating counterflow towers because there is less tendency to recirculate the hot moist discharge air.

CROSSFLOW TOWERS

Due to the configuration of this tower, whereby the entire sides are open air intakes, the higher elevations at the inlet areas tend to take a greater air volume into the tower. This creates a negative pressure zone close to the fan discharge which tends to encourage recirculation (Fig. 5). To this can be added the normal wind pressures and velocities. A recirculation situation can easily be set up. Since this problem is inherent in the design of a crossflow tower, it is difficult to overcome.

RECIRCULATION

Recirculation is one of the most serious problems in cooling tower siting. Some of the problems and solutions that substantially affect the capacity and efficiency of the cooling tower depend upon the amount of recirculation taking place.

When installing a large crossflow industrial tower, a little research should be performed into the direction and force of the prevailing winds. If it is determined that the wind is quite prevalent from a particular direction most of the time, a prudent selection to avoid recirculation would be to consider a single flow crossflow tower (Fig. 6). This should be located with the air intake louvers directly into the wind which minimizes recirculation. The added wind impact will increase chilled water temperature by approximately 1° for every 15 miles per hour (free air) of wind.

Fig. 3. Smaller air conditioning cooling towers sometimes are sited, for convenience, in enclosures or underground.

Fig. 4. Blow-through or forced draft cooling towers require a large area of free air intake and unobstructed discharge space.

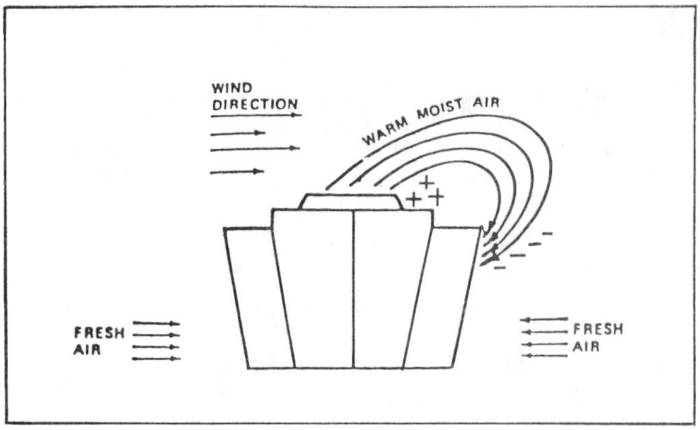

Fig. 5. Recirculation, whereby the hot moist exhaust air is run back into the tower through the air intake louvers is caused by negative pressure on many types of crossflow towers.

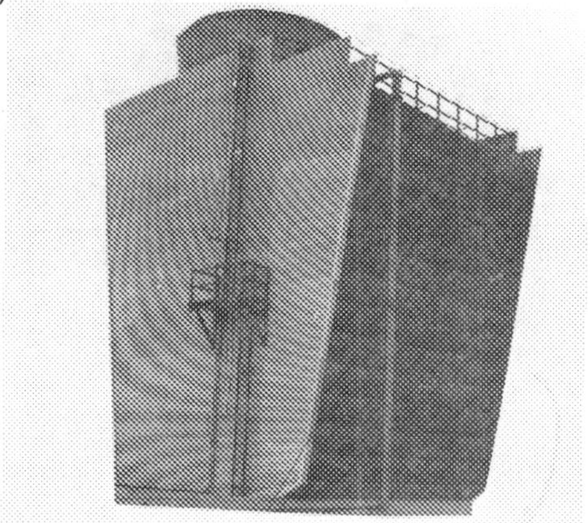

Fig. 6. A single flow crossflow tower can be erected facing the prevailing wind.

A double flow crossflow tower can be sited in one of three manners relative to the prevailing wind:

A. Perpendicular
B. At an angle.
C. Parallel

Location C with the blank wall facing the wind, will minimize recirculation. Location B with the angle into the wind, will be slightly better than A with the prevailing wind into the air intake louvers. (Fig. 7).

Forced draft towers placed underground for "architectural compatibility" is an extremely large waste of money. To obtain performance anywhere close to design conditions, a 3°F to 5°F increase in selection of wet bulb temperature must be considered since the recirculation factor will be quite high. Eventually, the owner will discover that the only solution will be to construct an exhaust duct about 10 feet high around the tower to prevent contaminating the fresh air intake area. Considerable static pressure is lost to get inside the enclosure and through the tower.

In the case of installing batteries of towers to handle large tonnages, starvation and recirculation will occur if walls and barriers are erected to attempt to hide these units. The installation of velocity regain cylinders from 10 feet to 22 feet high will go a long way in preventing recirculation in a counterflow tower and minimizing it in a crossflow tower.

In many cases, buildings are erected adjacent to towers which are higher than the fan cylinder discharge. Prevailing winds can create a recirculation pattern in counterflow or crossflow towers. However, this can be eliminated by installing elevated fan stacks. (Fig. 8).

Fig. 7. Crossflow towers, where possible, should be sited with the blank end wall facing the prevailing wind.

The use of Venturi section velocity recovery stacks provides an added bonus of increasing the air flow approximately 7 percent by reducing the exit pressure at the fan. This is due to Bernoulli's theorum which states that "when the velocity of a fluid is increased, it's pressure is decreased". Efficient air handling produces an improvement in cooling tower performance.

When a cooling tower is incorrectly sited so that the air inlet to the tower becomes choked or undersized, the following happens to the overall system:

1. The air volume available to the tower for cooling is decreased due to the increased static pressure across the fan.

2. More condenser water will be required, larger pumps and horsepower are needed which will escalate the energy costs.

3. With more water demanded for the tower, more water treatment would be required.

Specifying engineers and process engineers must intelligently consider siting a tower and resist the attempts to apply "aesthetics" to a cooling tower installation. For cooling towers to perform their jobs efficiently, adequate air must be provided and sufficient discharge areas must be available.

Fig. 8. In many instances, recirculation can be minimized by installation of a velocity regain FRP fan cylinder.

UPGRADING ECONOMIES

An existing tower, whether it be chemical process or building refrigeration, with problems producing enough cold water for the process, condensers, and compressors or when additional capacity is required because of expansion, should be carefully scrutinized by a specialist to determine the upgrading potential and cost versus the cost of installing new equipment. The engineering state of the art and modern day materials permit a wide latitude of upgrading to existing units.

Wood splash bar fill, (Fig. 9) when changed to cellular configuration, whether it be counterflow, crossflow or blow-through, will produce dramatic results in colder water. An actual case study of a blow-through tower such as is shown on Fig. 10. shows a dramatic improvement for minimum dollars as indicated by the following case.

CASE I

In this case, the existing corroded and blocked steel fill was replaced with PVC cellular fill. The 3880 small 1/8" diameter orifice nozzles (388 per tower for 10 units) were changed to 120 large 1¼" diameter opening non-clogging nozzles, (12 per tower for 10 units) on PVC pipes. Existing rusted steel drift eliminators were changed out to non-corroding PVC cellular configuration.

Replacement of wooden slat splash bars in large crossflow industrial towers with plastic V-Bar units, which are combination splash and film fill, have resulted in up to 50 percent improvement in tower performance. (Fig. 11).

Fig. 9. Towers with old-fashioned wood spash bar (Left) which eventually deteriorates, can be retrofitted with high efficiency cellular fill (Right) to maximize cold water.

Counterflow towers have been more than doubled in capacity by changing water distribution from troughs to low pressure spray and installing cellular fill in place of splash bars.

Crossflow towers have shown improvements by merely changing splash orifice nozzles to target orifice nozzles. (Fig. 11)

Each cooling tower installation must solve a particular problem and each should be a custom engineered design. To optimize the tower's operation and maximize the Corporate investment, selection responsibility should be distributed among the purchasing, operating, and engineering department, and the outside Consultant.

Improper siting of a cooling tower diminishes a portion of its effectiveness and may result in expenditures which do not optimize plant operations.

Fig. 10. Blow-through towers can produce at over 50 percent higher levels of efficiency by changing out the existing deteriorated corroded steel fill, piping and drift eliminators.

PVC "V" Fill Bar

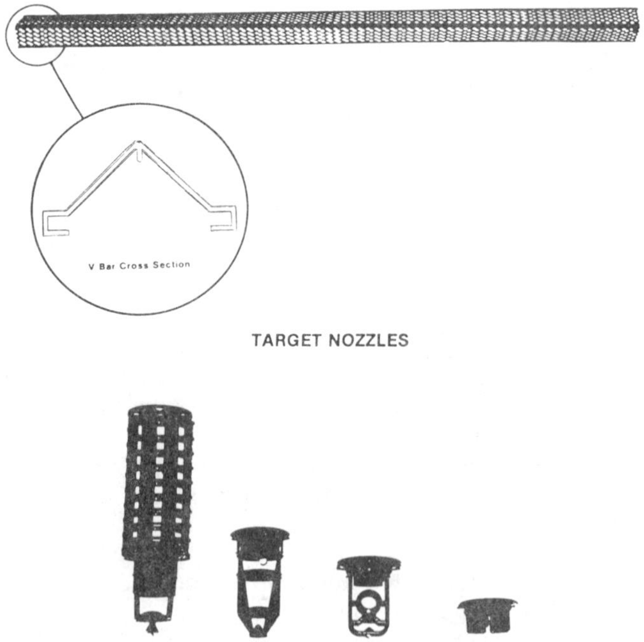

Fig.

Fig. 11. To maximize retrofit of large crossflow cooling towers on an economical basis, wood spash bars and grids can be changed to more efficient ledge-type perforated self extinguishing PVC and the water distribution orifices can be upgraded with target spash metered orifices.

CHAPTER 12

WINTER OPERATIONS

FIG. 1. The winter handywork of Jack Frost is beautiful and decorative in the proper framework and perspective.

For winter operation in below freezing temperatures the problem of icing is an ever present hazard that the cooling tower operator must be aware of. Severe icing has occurred as far South as Baton Rouge, Louisiana and in some areas of Texas during the winter, discharge plumes are known to blow over and freeze on roadways causing dangerous driving conditions.

The cost of ice damage can be quite high in replacement of wet decking fill, air intake louvers, and possible structural members. With proper control, icing damage can be eliminated, or at best, held to a minimum.

Where cold weather operation is normally expected, a source of heat is injected into the cooling tower and controls are installed to regulate the airflow. With normal water circulation, the basin should not freeze, but emersion electric heaters or steam coils are activated as precautionary measures.

FIG. 2. Winter operations of cooling towers present many problems. However with the proper understanding of the principles involved, hazards can be avoided.

One of the main winter operational controlling procedures is airflow management. Automatic control systems utilizing temperature sensitive probes can be installed in the sump of the basin to be set automatically to reduce fan speeds. This, of course, is only available on two speed motor configurations. The purpose of reducing the fan speed is to permit the hot water to travel throught the wet decking fill allowing the ambient cold air to cool the water to the desired temperature. Depending upon the control setting and process requirements, the fans can be completely turned off in frigid weather, thus maintaining higher temperatures within the tower to assist in preventing

FIG. 3. A major cause of air intake louver icing is broken or out of position splash boards. The ice must be removed physically and reverse fan heat applied periodically to prevent build up.

icing. While the reduction of airflow is cut back considerably, the second principle of icing control to bear in mind is that full water flow should be maintained at all times. In winter multi-cell operation of cooling towers, it might seem logical to some operators to spread the load amongst all cells. However, the opposite is the desired and recommended procedure, namely concentrate the water load on one or two cells, completely shutting off the excess capacity.

FIG. 4. Internal damage to wet decking fill and supports can be substantial requiring expensive repairs.

It is important that all of the flow control valves be operational and completely turn off the water to the inoperative cells since a small trickle of water could be disastrous. This would very rapidly accummulate ice, freeze solid, and cause expensive damage. In other words, a percentage reduction of water flow over one cell only intensifies the icing problem.

When ambient weather is below freezing, the effective control can be accomplished by variations in the fan speed. During operations in the winter, return water from the tower is quite lower than that required for efficient plant operations during higher ambient temperatures in the spring, summer and fall. Therefore, full airflow and water flow to all of the cells is not required to maintain production levels. By reducing the airflow substantially, this lowers the heat transfer rate in the cells and the operating temperature of the water inside will rise. This higher temperature inside the cooling tower will assist in preventing ice formation.

The tower must be monitored by the operator constantly during these critical temperatures and usually every two to three hours the fans should be rotated in reverse mode for 15 to 20 minutes. Caution is advised that if the fans are operated in reverse for more than a half hour, ice could form on the fan stacks, fan blades, or drift eliminators causing hazardous and dangerous conditions.

Installation of automatic variable pitch (AVP) fans are a viable option for energy conservation and icing control. These AVP fans have built in control circuits that sense the load automatically and adjusts to the most economical operating load. The fan can be cycled to reverse the air flow automatically for periods designated by the operator. During severe icing conditions, the AVP or alternate control systems must be carefully monitored to prevent malfunctions.

FIG. 5. Automatic Variable Pitch (AVP) fans, besides being energy savers by sensing heat load, can also be automatically cycled to reverse pitch to melt ice build-ups.

Splash deflectors inside the cooling towers located at the inside top of the air intake louvers in counterflow towers and at the level beneath the hot water basins in crossflow towers should be inspected before winter operations for proper maintenance. If these areas freeze up, they can reduce the airflow through the tower considerably and block air movement as the ice builds up.

Crossflow towers present problems that are more severe than counterflow. Counterflow icing is easily controlled by reverse fan operation or installation of bypass water distribution which sprays the water beneath the wet decking fill, bypassing the wet decking area thereby eliminating the icing problem in the fill, and since ambient temperatures are low enough, the counterflow tower becomes an atmospheric spray chamber. The spray nozzles and distribution arms can be so designed to be a considerable distance away from the air intake louvers to prevent air intake louver icing. A partial solution in crossflow towers where ambient temperatures are very low would be to bypass the top hot water distribution basins and pump the hot water directly into the basin so that it will act as a cooling pond.

Again, it is important to mention that during winter operations, the heat and water load is not as high as warm weather operations and the water load must be concentrated in as few cells as possible for full load and not thinly distribute the water throughout the tower. The low volume of water will freeze up rapidly causing severe damage in crossflow towers. The lowest temperatures are at the air intake louver face at the lower section of the fill.

FIG. 6. Ice damaged fill must be replaced in order to obtain proper heat transfer from the wet decking.

If the pumping volumes are low, bypass valves should be activated to put all of the water into one cell and as an assist to keep the water away from the air intake louver edge, these holes can be temporarily plugged up with rubber laboratory corks. (Be sure to remove these obstructions during full volume operation.) In this procedure however, monitor for over pumping which will "spill" water over the air intake louvers and the leading edge of the face area of the fill which will set up icing conditions.

When the cells must be shut off, the last cell of the crossflow tower must be operated to prevent freezing of the water at the end of the distribution header pipe. When there is the smallest heat load on the tower during startup and shutdown, the operator must carefully monitor the tower since this is when it is vulnerable to icing damage. The water bypass would be valuable to provide good cold weather control. By this means, the water can be circulated throughout the plant bypassing the freezing hazard of the wet decking and air intake louvers. The bypass can be used until the water temperature rises above 80 deg F and then controlled by maintaining the operating temperatures by partially opening the valve. The fans should not be activated until the circulating water is in the area of 80 deg F and then monitored to determine the speed.

During cold weather startup, the bypass should be activated until the 80 deg F temperature is reached. At that point the full flow of water should be put over the fill without fan operation. As temperatures rise,

FIG. 7. An advantage of cellular conversion in crossflow towers is that the "danger area" between the fill and the air intake louvers can be widened preventing falling ice damage.

the fan rotation is controlled as previously discussed. If the crossflow tower is not equipped with a bypass in the circulating water system, the pumps are to be activated at the last possible minute. When the circulating water reaches 80 deg F add the fans one at a time as the heat load increases. Increase the heat load as fast as possible until the recommended minimum operating temperatures are reached. During shutdown, the procedure is reversed.

After exposure to icing conditions, the operating personnel should inspect the louver blade areas, lower fill sections, and structural members for damage and necessary repairs be scheduled and effected.

With careful operating procedures and maintenance, cooling towers can survive the winter icing experience.

FIG. 8. Counterflow icing can be minimized by installing the wet decking cellular fill above the air intake louver openings. The enclosed plenum chamber within the counterflow tower walls are more condusive to reverse air flow to contain heat preventing ice build up.

WINTER OPERATION

A. GENERAL DISCUSSION

1. When ambient temperature is 40°F and falling, winter anti-icing operations must be implemented.
2. Ice formation should be prevented if at all possible, since severe damage to the interior components can occur.
3. The most sensitive areas for icing to form are the closest to the air inlet louvers and the tower face exposed to the wind will be most prone to freezing.

B. WINTER OPERATION INSTRUCTIONS

1. Counterflow towers have less tendency to ice up then most crossflow towers of this capacity due to the fact that the wet decking fill is removed from the air intake louvers in counterflow where most damage occurs from ice falling off the louvers in crossflow configuration.
   a. This fill arrangement allows for warmer air at the air inlet without overly relying on air flow characteristics.
   b. Deicing strategy is to raise the temperature of the water passing through the fill by varying the number of fans operating.
   c. If extreme conditions such as reduced heat load exist, the tower should be operated in the bypass mode.
2. Normal winter operations.
   a. Check that all de-silting and drain valves are closed.
   b. Check that the bypass valves are closed.
   c. Check that all cell shutoff valves that are operating are fully open.
   d. When ambient temperature approaches 32°F (0°C) monitor hot water temperature.
   e. Refer to Winter Operations Chart in freezing weather to determine whether tower is operating on that drawing.

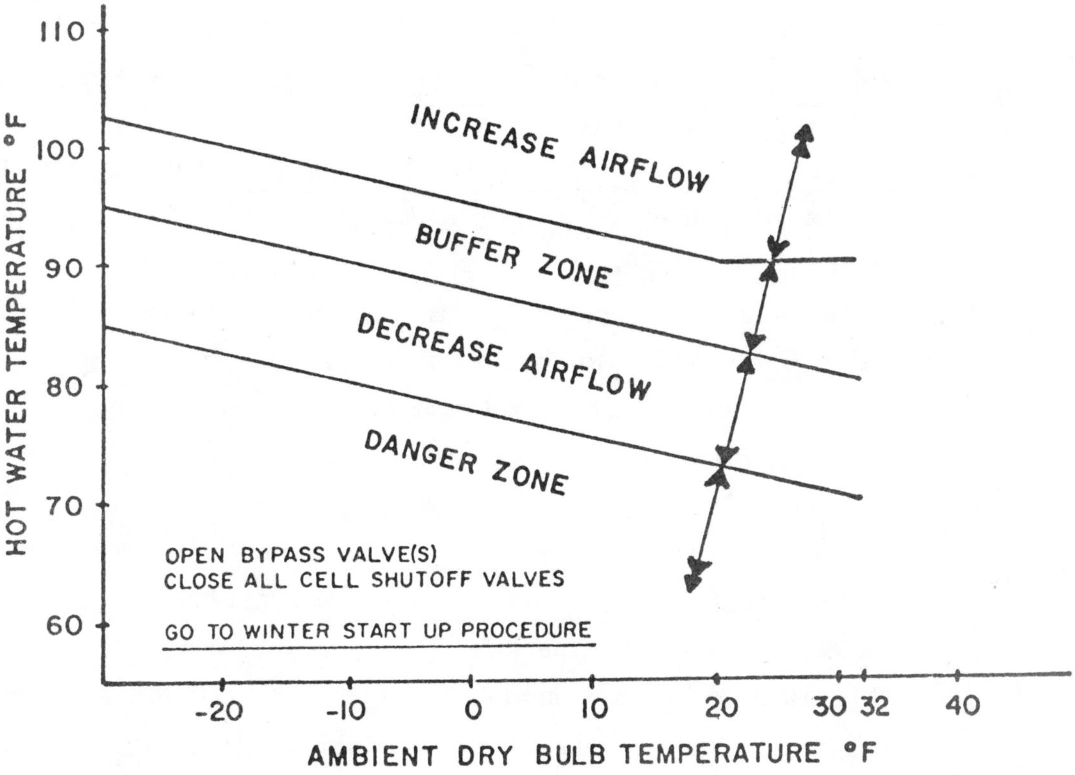

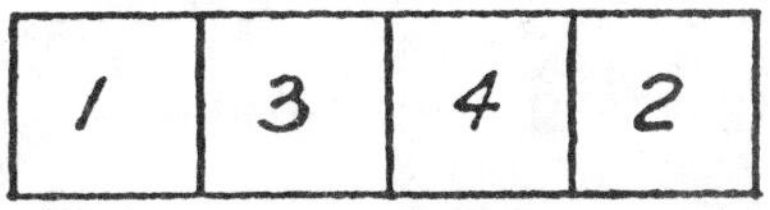

TURN FANS ON IN ORDER SHOWN
TURN FANS OFF IN OPPOSITE ORDER

i. Buffer zone - tower remains as "Is". Monitor ambient and hot water temperatures.

ii. Decrease air flow region - turn fans off one at a time as indicated by sequence, allowing 5 minutes between each fan stop for the water temperature to react.

iii. Danger zone region - turn all fans off immediately, increase heat load or flow on the tower to maximum available. Final step is to operate in bypass mode.

iv. Increase air flow region - turn fans on one at a time in order listed allowing 5 minutes between each fan start for water temperature to react until all fans are on.

v. Reverse fans - if capability is available, run fans in reverse a maximum of 5 minutes per hour to blow warm air down through the tower. If gear box lubrication system does not operate in reverse, maximum 5 minutes should be strictly adhered.

vi. Operating in bypass mode, frost due to condensation can buildup in fill. This can be eliminated by putting water into the tower fill for a short period of time.

vii. Maximum heat load and water volume should be sent through wet decking fill. Reducing water below Design will enhance freezing the tower.

C. WINTER START-UP

1. Winter start-up is defined as any start-up when the ambient air is at or below 32°F. The listed procedure is to be followed: -

   a. Check that de-silting and drain valves are closed.
   b. Check that all cell shutoff valves are open.
   c. If necessary, fill the basin and pump suction system.
   d. Check bypass valves open to proper position.
   e. Start bypass operation.
   f. Monitor hot water temperature.
   g. As load increases, temperature will increase.
   h. When hot water temperature exceeds 90°F, close bypass.

I. Tower is now in normal winter operation. Refer to previous section on normal winter operation.

D. <u>WINTER SHUT-DOWN</u>

1. Place tower in bypass mode (bypass valves open, flow reduced).
2. Flumes and asbestos-cement distribution pipe should drain automatically.
3. The basin and bottom part of the risers have to be protected from freezing by draining or adding anti-freeze.
4. If outage is of short duration and/or freezing is not very severe, it may be sufficient to keep a small amount of warm water circulating through the bypass to prevent freezing.
5. If the tower is shut down for a long duration and or freezing is very severe, then the basin should be drained if anti-freeze methods cannot be implemented.

CHAPTER 13

MAINTENANCE

In the pressured course of daily events, MAINTENANCE many times is relegated to a second class citizens status. Over the years, unfortunately, Murphy's prime law, *Things left to themselves tend only to get worse,* has been expensively proven to be correct. Maintenance and inspection is an on-going process and must be programmed and implemented.

We will cross that bridge when we come to it has no place in the Maintenance Manager's lexicon. That "bridge" is here and now.

FIG. 1 — Destructive corrosion necessitates steel replacement.

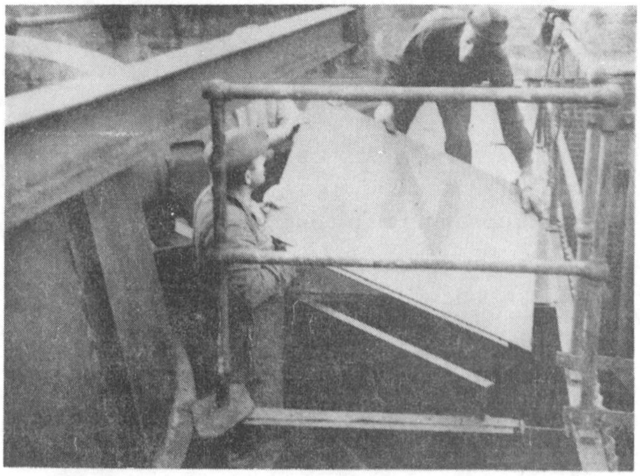

FIG. 2 — Rebuilding roof deck destroyed by corrosion.

FIG. 3 — Acetylene torch is used to weld replacement steel when deterioration becomes excessive.

FIG. 4 — Chipping tool is used to remove rust and scale. A little corrosion goes a long (and costly) way inadequately treated.

COOLING TOWER MAINTENANCE AND REPAIR

Cooling tower corrosion is an item of maintenance too often overlooked by building and plant management personnel. Purpose of this four-part series will be to alert responsible personnel to the need for a continuing program and to review pertinent maintenance procedures. Corrosion principles are discussed first, to be followed by techniques of cleaning, sandblasting, repairs and maintenance. The concluding article will concern methods of increasing cooling tower capacity to obtain more effective cooling and handling greater tonnage of refrigeration with existing equipment or with minimum expenditure.

Compressors and evaporators of an air conditioning installation are the pampered darlings of the system. At the end of the air conditioning season off come the heads, and the machines are carefully overhauled and made ready for next spring. If anything goes wrong during the operating season, experts may be called in to remedy the situation.

Basically, there is no complaint with this procedure. Equipment in the basement or engine room is quite expensive and must be adequately maintained.

However, the third member of the air conditioning team

– the cooling tower, whose presence, in good operating order is just as necessary – is frequently the orphan of the maintenance department and, invariably, is left out of the operating budget by the building owner or real estate management.

These omissions are not deliberately; there are many reasons, among them:

(1) Cursory inspection of the corrosion protective coating indicates "all is okay" . . . or, there are a few spots, "we can knock the rust off, and paint over it".

(2) The old adage "out of sight, out of mind" applies to a degree. Most cooling towers are perched atop the building, since many towers were built after the structure was erected, usually located in an inaccessible location, since they are not income producing items. Location can make routine maintenance difficult.

(3) There are not enough men in the maintenance department to spend enough time (on the roof in cold weather) to do a thorough job, which should include taking out, and later properly replacing the wood fill and drift eliminators.

(4) It hasn't been until recently that Polymer Chemistry developed adequate coatings specifically designed for cooling tower problems and operating conditions.

Now, after what appears to have been years of trouble-free operation owners find they are faced with extensive overhaul and replacement.

This is a sobering thought, because approximate replacement cost in the New York metropolitan area is about $55 per ton of refrigeration. This involves a great deal of engineering to custom fit the new tower to the system, including plumbing and electrical work, building department surveys and permits to utilize dunnage beams, plus demolition and removal of the deteriorated tower.

Simple arithmetic should convince you of the value of adequately maintaining your cooling tower.

Primary interest of the chief engineer and building superintendent is: "How can I prevent this from happening to MY equipment?"

Answer, of course, lies in proper protection of steel surfaces against the highly corrosive atmosphere in which we live. Oxidation, ultra-violet ray degradation, sulphuric acid attack must be protected against. Not to be forgotten is expansion and contraction which the coating must flex without losing vapor-tight integrity.

Many chiefs have their men earnestly attack rust and corrosion with chipping hammers, wire brushes and scrapers, and with much zeal; then prime and coat exposed steel.

This is commendable, But it only temporarily halts the corrosion process. Rust and corrosion are more than a chemical process, ie, oxidation of steel by exposing metal to oxygen and moisture. Unless corrosion and corrosion products are removed completely, like a decay in a tooth the process will continue.

Ferrous corrosion, as explained by the National Association of Corrosive Engineers, is the dissolution of iron in water . . . essentially an electrochemical process. Iron dissolves at active points on the metal surface called anodes, leaving electrons in the metal. To maintain neutrality in the system, these electrons flow through the metal to other points called cathodes, where reactions which consume electrons can occur (Fig. 6).

Corrosion points – places where anodes develop in a system – have a number of identifiable sources: rivet or welded joints, dissimilar metals, oxide scales or rust pockets, differences in metal miscrostructure, etc.

Total corrosion reaction is a combination of separate anodic and cathodic reactions, depending upon each other, and controlled by the slower of the two reactions.

Absence of oxygen, such as heavy coating over hand-cleaned rusted steel will not stop this electrochemical reac-

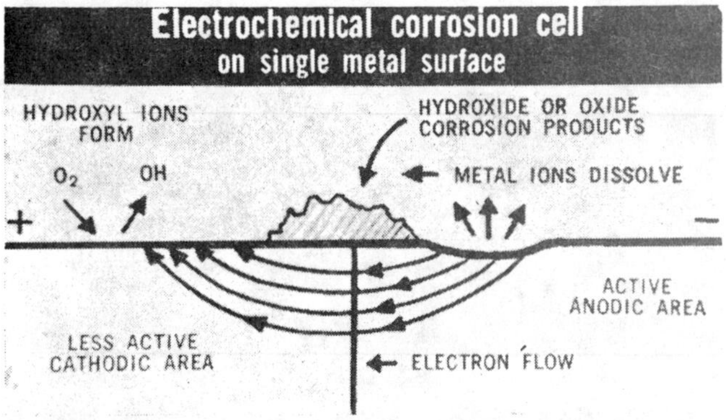

FIG. 6

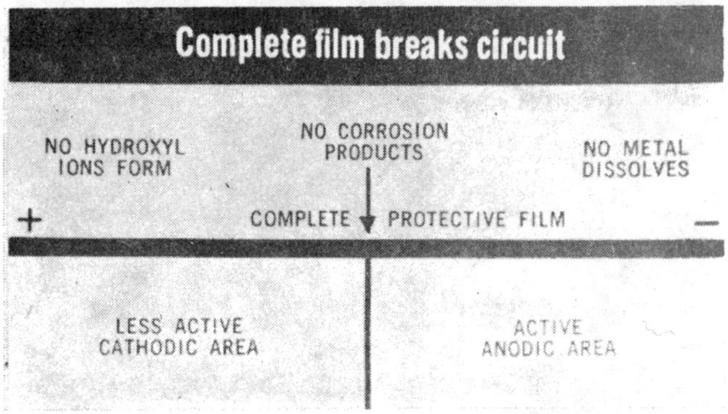

FIG. 7

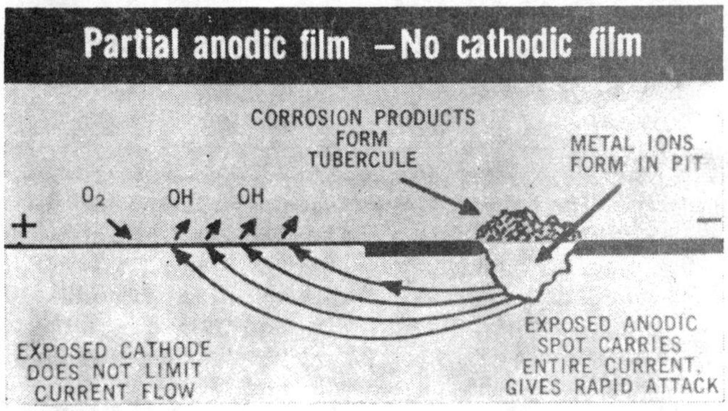

FIG. 8

tion. It will only slow the process down in direct relationship to the percentage of corrosion left in the "valleys" of the steel microstructure. Only complete cleaning . . . sandblasting, acid dipping or cathodic protection (sacrificial zinc) – can halt and control corrosion. (These methods, along with various hand cleaning alternates, will be discussed in a later article.)

A complete film or protective coating breaks the corrosion reaction by preventing the water of chemical liquid (electrolyte) from contacting the metal (Fig. 7). A break in the coating caused by mechanical abrasion, chemical attack of the liquid on an improperly selected coating material, or a pocket of corrosion under the coating caused by mill scale or inadequate cleaning, will concentrate the entire corrosive attack on the exposed anodic site, resulting in severe pitting (Fig. 8), attack on the coating, and further spread of the corrosion cell.

Therefore, the two main principles of corrosion protection by film coating of steel are:

(1) Adequate surface preparation – and

(2) Proper selection of protective coating for cooling tower exposure.

WINTER WEATHER MEANS cooling tower repairs. Due to improper coating (**Fig. 1**), roof deck and fan cylinder sections are being replaced and interior of tower is being sandblasted to remove corrosion. **Fig. 2** shows Butyl Isoprene Rubber being applied in cold weather. Note "rubbery" texture of the material. **In Fig. 3**, Epoxy and Fiberglass cloth patching are being used instead of expensive replacement of fan cylinder. Note holes in motor panel. **In Fig. 4**, basin is being protected with Epoxy and Fiberglass cloth. Steel was acid dipped after fabrication, shop primed, hand cleaned after erection, and reprimed before coating.

New Developments in Modern Coatings

The tremendous advance made in the field of paint and coating technology has resulted in a great variety of new protective coatings for steel structures — particularly in coating cooling towers and lining steel water storage tanks. Cooling towers require specifically designed materials, as they are subjected to severe environmental conditions — water immersion, chemical and atmospheric contamination, heat, freezing cold, mechanical vibration, and abrasion. The combined efforts of many people in the protective coatings industry, and many reputable manufacturers of paints and coatings have resulted in the formulation of improved methods and products.

This article intends neither to present a discussion of paint chemistry nor to try to assess the advantages and disadvantages of all coating systems, but rather to assemble certain basic information acquired by the author and his organization the hard way — testing and field application under operating conditions. The "hard way" refers to certain experiences obtained through the costly procedure of a particular coating system failing within a contractual guarantee period, and our repairing it at no cost to the purchaser (Fig. 1).

There are countless applications where one specific coating will be highly superior to all others — including the myriad of materials not listed in this article, such as zinc rich primers, stainless steel and aluminum pigmented resins, Polyvinylchlorides (P.V.C. Vinyls), Polyurethanes, Furanes, Teflon, Tar-Epoxies, ad infinitum. However, as explained before, this article is devoted to our experience within the cooling tower protective coating field. With this understanding, we list the following coatings and paint materials in what we have found to be their order of performance:

1 — BUTYL ISOPRENE RUBBER: Butyl has perfect properties for cooling tower use — flexibility, resistance, impermeability, ease of application directly from the can with no additives other than thinners — by brush, roller or spray. Two coats by brushing or rolling can build the protective film up to 12 to 16 mils dry film thickness — three spray coats properly applied will build up the same film thickness without sagging. Butyl Rubber is self vulcanizing at ambient temperatures and can be applied in below freezing weather at an overall moderate cost.

One of the most important factors to be considered in the selection of a protective coating for cooling tower rehabilitation is to be noted in **Figs. 1** and **2** . . . . repairs in the northern and central sections of the country must be effected during the part of the year when the tower is turned off and drained. This, by necessity, is during the winter, and Butyl Rubber is the only material tested that can be satisfactorily applied in below 40°F temperatures that will adequately protect the cooling tower steel.

Butyl will not soften and flow in extremely hot weather, nor will it embrittle and crack in below freezing temperatures. For repair work, it welds to old coatings of Butyl after washing down with thinners. This material is used in conjunction with a Phenolic based red lead rust inhibiting primer system. Due to its high density and impermeability to gases and moisture, it takes considerable time for Butyl to set. It is a solvent evaporation and release curing system, and the speed of the cure is a relative function of the temperature.

At extremely low temperatures, once Butyl is applied, it forms a protective film, and gradually cures. At warmer temperatures, it sets faster. In 80°F weather, Butyl cures within 12 hours. It is expected that this material will take longer drying time than ordinary thin film coatings since a minimum of 10 mils is recommended for adequate protection of the steel, because for complete cure the solvent must migrate through the coating. The high solids content of Butyl permits heavier coats with a minimum expenditure of labor.

2 – HYPALON: A DuPont synthetic rubber, made by reacting polyethylene with chlorine and sulphur, producing an ambient temperature vulcanizable rubber. The polyethylene rubber in combination with equally resistant pigments produces a tough resilient coating that has outstanding resistance to acids, weathering, abrasion, and various chemicals. Since it does not contain any plasticizers, the rubber itself is its own permanent plasticizer; it does not dry hard and brittle as do standard paints and unsaturated materials that oxidize over a period of time and begin flaking and brittling.

Hypalon has a high solids content and can be built up in three coats, (rather that 4 to 6 as with other materials) to a 10 to 15 mil dry film thickness which will afford maximum protection with a minimum labor cost. For patching, the liquid Polyethylene will weld to itself with proper cleaning of the base coat first. The only drawback to using Hypalon in cooling tower work is that the manufacturer does not recommend its use in temperatures below 45°F, and the majority of cooling tower coating must be done during the winter when this critical temperature cannot be controlled.

3 – LIQUID NEOPRENE: A heavy duty resistant chemical material formulated from synthetic rubber polymers. The best protection is afforded by a two-part mixture of accelerator and vehicle. This must be handled carefully since once a mixture is made, it will cure rapidly. The surfaces should be very clean (white meal sandblasted) for maximum results. It is recommended that Neoprene be used on new installations where the steel can be acid dipped, shop primed, and field coated. To apply Neoprene over old surfaces, the applicator should have considerable experience and the metal should be completely sandblasted free of corrosion and the old coating. A minimum of four coats (3 to 4 mils maximum thickness each per coat) is recommended to build up adequate protective film on steel surfaces.

Neoprene is highly resistant to all corrosive elements encountered in cooling towers, and if properly and carefully applied, should give many, years of trouble-free protection. However, like Hypalon, temperature is critical. Neoprene cures by chemical catalyzation, and the chemical reaction becomes sluggish at 45°F and practically halts at 35°F and below, leaving the surfaces unprotected until the temperature rises.

4 – CATALYZED EPOXY: This is a highly resistant modern chemical coating. High resistant Epoxies are supplied in a two-part mixture, and caution should be taken after the catalyst is added since the pot life is then rather short. Care should be exercised in handling this material since the strong components could cause severe skin eruptions. Manufacturer's instructions should be closely followed. Epoxies, for best results in film coating steel surfaces, should be applied over well prepared surfaces. They dry to a very hard finish, and due to an extreme hardness, tend to be rather inflexible.

Generally, Epoxies are not recommended for use on basins without reenforcing with Fiberglass cloth to absorb the movement of the steel. Epoxies are used to good advantage in repair work in conjunction with Fiberglass cloth. As the **Fig. 3** shows, expensive steel replacement can be avoided with the proper application of Epoxy. For this use, Epoxies can be especially formulated to cure in colder temperatures.

Epoxy, when applied outdoors during temperatures above 45°F affords excellent protection to steel surfaces. Fiberglass cloth and Epoxy is being applied to the basin, interior flanges, and sumps of the cooling tower in **Fig. 4.** This is a new installation under construction, the steel was acid dipped and factory primed, and the work was performed during the month of July. After five years of service, it is holding up very well.

FIG. 5 — Photo shows pile of bitumen coating from cooling tower after about five years of application.

The most striking quality of the Epoxy family of coatings is their tremendous adhesion. Like Neoprene, this is a chemically catalyzed material, and manufacturer's recommendations should be followed. Most Epoxies should be applied in over 45°F temperatures. A good grade Epoxy will cost more than the other materials. However, only five mils of Epoxy over steel, sandblasted to white metal, properly primed, applied within ambient temperature limitations for adequate cure, will give the required protection against corrosion, while the other Polymer coatings require a much heavier application to give the same protection.

5 – ASPHALTUM: This material has many trade names such as Bitumastic, Insulmastic, Gilsonite, Tar, Maltha, etc. Basically it is manufactured from a bitumen, as a by-product of coke production, and oil refining, and consists mainly of hydrocarbons. Bitumen is the generic family name of the substance, and it is applied hot with mops, cold with mops or brushes, or sprayed under pressure together with a mixture of Asbestos fibers, as in automobile undercoatings, to an eighth to a ¼" thickness.

Since the Bitumen is not in itself attacked by rust or corrosion, this can lead to serious problems since once corrosion starts beneath this thick mastic covering (which generally has moderate to poor adhesion to steel) it cannot be detected until damage becomes apparent in the steel structure.

This material is an old standby which has lost favor with modern corrosion engineers for steel protection due to certain continuing drawbacks. Since it has no flexibility, it loses adhesion, and because the cohesive strength is greater than its adhesive strength, it flakes off in sections when applied to vertical or overhead surfaces. The material crazes and cracks after the evaporation of the plasticizer. Microscopic hair cracks develop when the steel flexes causing moisture penetration beneath the usually heavy coating. This is insidious because there is no indication from the appearance of the coating that deterioration is occurring until the pitch falls off leaving the corroded and structurally weakened metal.

Fig. 5 shows typical results in a heavy coating of bitumen as applied on cooling tower steel. The basin in some areas was covered with as much as ½" thick. Most of the beams were originally covered with up to ¼" thickness of tar. A heavy blow with a hammer during cold weather shattered most of the material from the beams; a scrapping iron and chipping hammer took care of the rest. The basin was easily stripped with pneumatic chippers. This "pile" was in the cooling tower less than five years. Extensive sandblasting was required and new steel braces had to be added to the gear box and fan supports before recoating the structure with Butyl Rubber.

6 – ORGANIC OILS AND ESTERS: Due to their chemical structure, these materials do not have the basic resistance to stand up to normal operating conditions encountered in cooling tower operations. Most organic oils and esters are not impermeable enough to moisture and cannot be built up in thicknesses of any consequence to protect metal surfaces, especially under constant water immersion. For example, one coat of Butyl Rubber can be applied up to eight mils dry thickness without sagging or running, whereas oils and esters can be applied to about two mils per coat (one mil is equal to one thousandth of an inch). The organic oils and esters have a tendency to brittle out as the polymer oxidizes. Some paints, especially those based upon fish oil, saponify in sulphuric atmospheres (in most large cities) and would fail rapidly once this process starts. Organic oils and esters are not recommended for interior cooling tower applications to protect against corrosion.

7 – SURFACE PREPARATION: The best possible coating system, skillfully applied, can be of little value unless the steel surfaces are properly cleaned and prepared. It is impossible to overemphasize the dependence of long coating life – and thus economy – on adequate surface preparation. This important area, surface preparation and application, will be the subject of Part III.

How to Clean/Coat/Repair Your Cooling Tower

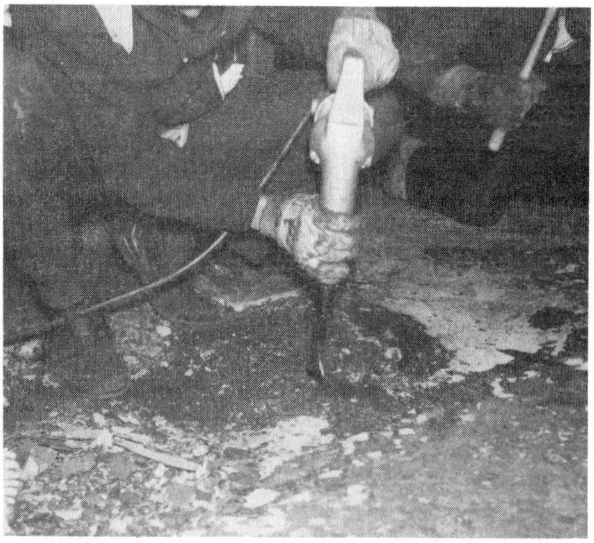

FIG. 1 — Heavy scale and coating being removed by mechanical chipper preparatory to sandblasting.

FIG. 2 — Sandblasting large cooling tower. Note protective hood for workman, and plysheets.

We live in an extremely corrosive atmosphere and environment, engaged in a never-ending fight against the forces of nature. Given sufficient time, corrosion and erosion can reduce man's works to rubble. From the moment steel is refined nature relentlessly seeks to return it to its original oxide state.

Thus, the corrosion engineer is vitally interested in answering two questions. . . (1) what is the best method to halt corrosion, and (2) what is the best protection to prevent further corrosion. (As we are concerned here with cooling towers, our prime consideration will be ferrous corrosion).

Where conditions permit, sandblasting is the desired procedure for removing rust. On this, there is no doubt among the experts. Their thinking is oriented to methods and techniques of sandblast cleaning rather than arguments for or against.

Few commercial, residential buildings and industrial plants have sandblast equipment and crews, so this work is contracted. The following are a few points you as the

FIG. 3 — Interior steel of cooling tower partially sandblasted. Note contrast between blasted, non-blasted areas.

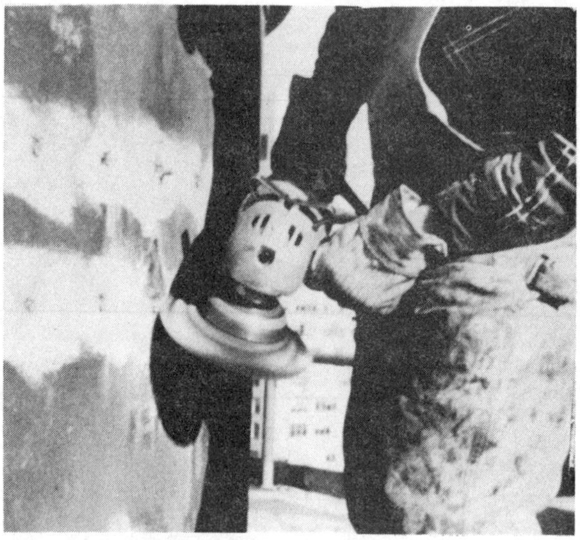

FIG. 4 — Mechanically wirebrushing light corrosion on exterior.

owner's representative should consider to insure that the job will be accomplished as agree:

1 - Reputation and Experience. If possible, obtain a recommendation from a fellow engineer for whom the contractor has done work. Visit and inspect a similar installation. A reputable contractor should provide a 5-year maintenance guarantee. Ascertain that your contractor has been doing this type of work for at least five years, and is not using our installation for "on the job training."

2 - Equipment - Air pressure and air volume (plus size of the blast nozzle orifice) propel the abrasive material to sandblast steel to white metal. Recommended minimum pressure is 100 psi with a volume of 100 cfm of air. Horsepower to generate these requirements depends on the diameter of the nozzle bore. A ¼″ nozzle requires 18.1 HP to produce 103 psi at 101 cfm, while a ⅜″ bore needs 20.0 HP to turn the equivalent, both using full grain abrasive size. Should the contractor show up with a 5 or 10 HP compressor, be assured that this is inadequate. A minimum of 20 HP electric is needed, or a 165 cfm gas or diesel rig, since allowances must be made for pressure drop between compressor, sandblast generator, and blast nozzle at the work area.

3 - Work Method - Before blasting heavy corrosion and old coatings of mastic must be chipped and scraped, otherwise the abrasive will bounce off and require excessive pressures to clean. Mechanical chippers, hand scrapers and hammers can do the job, as shown in **Fig. 1.** After preliminary chipping, blasting should proceed from top to bottom.

Note that only those areas should be cleaned which can be primed the same day. Do not leave sandblasted surfaces unprimed over night, since differences in day-night temperatures cause sufficient moisture condensation on bare steel to start an invisible film of rust.

During sandblasting, plywood sheets and canvas drop-cloths should be used to contain the abrasive grit (see **Fig. 2**). Operations should be suspended in relative humidities of over 85%.

Fig. 3 shows a panel partially sandblasted and clearly demonstrates the superiority of this method of cleaning. The magnified surface of corroded steel has the appearance of mountainous terrain with high peaks and low valleys. Blast grit is driven into every crevice and crack, cleaning down to white metal without leaving pockets of corrosion.

Since the prime source of corrosion is an electrolytic cell (as discussed in a previous article) and if all corrosion products are not completely removed, the entire corrosion process starts over.

Most coatings, if applied heavy enough, look good for at least a year. Basic aim of a competent contractor is to have his jobs last at least 5 years without necessary touchups or complete failure.

The success in halting corrosion depends on the cleanliness of the steel. The better it is cleaned, the longer the coating will last. . .and the longer the structure will last. Time and again this has been proved. On many occasions, I have seen metal where the corrosion in areas of thin dividing walls between cells had "eaten" through the steel. After blasting, priming and coating with Butyl-Tect, Butyl Isoprene Rubber the holes remained but the corrosion was completely stopped for periods of four years and longer.

Electrical mechanical wire brushes can be used on exterior steel surfaces where rust and corrosion are not excessive (**Fig. 4**). You may have the crew to do this job. However, all areas must be completely cleaned - bolt heads, welded seams and joints - since the life of the coating depends directly on the degree of cleaning.

Other mechanical means of cleaning include "Bumblebees", small, spiked rotating wheels that vigorously abraid metallic surfaces - and needle guns, ⅛″ diameter hardened metal rods that vibrate vertically in a holder which fractures the scale. There are other devices.

Hand wirebrushes, hand scraping irons and chipping hammers are old standbys. They require "muscle and sweat", however, and the results depend on the vigor and willingness of the workman. It is easy to prime and coat over rust, like sweeping dirt under the carpet. This only means that the job must be done again within a short time, with the resultant loss of metal thickness and strength.

Before sandblasting, the optimum paint system must be determined by conditions, whether it be Butyl-Tect, Butyl Isoprene rubber, Neoprene, Hypalon, Polyvinyl-chloride, etc. This insures that a compatible primer is on hand for daily application after blasting. Most primers dry within a few hours, very essential, so that sandblasting the following day will not disturb newly-primed areas or adhere to the primer.

Following sandblasting and priming, the tower is cleaned by blow down with air blast (minus abrasive grit) followed by a once-over with an industrial vacuum cleaner.

FIG. 5 — Roller coating Hypalon in multiple coats of contrasting colors.

FIG. 6 — Painting Butyl Rubber on interior steel structure, in multiple coats of contrasting colors.

The economics of maintenance coating and painting resolves around the simple basics of cost. Often, the difficult task is convincing management and then obtaining enough funds to do the job properly. Cutting expenses by doing an initial job of hand wirebrushing instead of blast cleaning turns out to be the more expensive route when the job has to be redone in a comparatively short time.

An excellent article on industrial and plant maintenance notes that the cost of sandblasting original work at the plant site is no more than the cost of hand cleaning commonly used in repainting, since a coat of fresh paint can hide a lot of trouble, but not for long. The article states: A two part catalyzed Epoxy system applied over mill scale with a minimum surface cleaning costs about 50¢ sq. ft. with a service life of one year (mill scale is a microscopically broken coating of corroded iron oxide). When applied to sandblasted steel, this system costs about 80¢ sq. ft., and has a service life of 7 years. This breaks down to an annual cost for protection of only 11¢ sq. ft. This figure is even less if spot repair maintenance is properly conducted. Failure to remove mill scale by blasting, in this example, wastes 39¢ sq. ft/yr, and with a plant having more than 3 million sq. ft. of steel requiring paint protection, the better buy is 7 years for 80¢, not one year for 50¢.

In view of the above, it is obvious that if the engineer can convince management to spend the higher initial cost and amortize it over the effective life of the protective job, he is much further ahead in dollars and man hours.

In a discussion at a meeting of the National Association of Corrosion Engineers, the maintenance engineer of Firestone Tire & Rubber Co. stated that his company is conscious of the fact that surface preparation represents 65% or more of the overall cost of the job.

A pre-planned program of inspection and repair is far less expensive than accepting cooling tower failure in coatings, wood members and/or mechanical equipment. Periodic maintenance can be relegated to weekends or other non-productive hours, eliminating downtime. Draining and inspecting can be done at regular intervals. Following visual inspection, testing be done to discover weak spots not visible to the naked eye. Adequate cleaning, priming and repairs should be made by trained personnel.

Knowledge of coating alone is not sufficient in maintaining cooling towers. Wood fill will amount to a veritable red wood forest on the roof. A common area of failure is the structure covered by wood fill and drift eliminators. This is often neglected because of the difficulty of removing wood fill, storing, repairing, and replacing to maintain a static pressure and design throughout as intended by the manufacturer.

Half measures are valueless when dealing with destructive corrosion. Fill must be removed, surfaces cleaned (sandblasted if neglected too long) and a good resistant coating properly selected for conditions, and properly applied. There are no short cuts to long range cooling tower maintenance.

An understanding of cathodic protection further emphasizes the electrolytic nature of corrosion and why it is so essential that all corrosion cells be removed . . . and that steel be coated with a heavy material for protection against attack from under the coating or from without. Electrolytic corrosion is nature's chief weapon in destroying metal. This process is somewhat like the action of a storage battery. Certain areas on the surface of steel tend to become anodic, others cathodic. Moisture, laden with ion-containing impurities, provides the natural electrolyte, and oxygen, which promotes the corrosion process.

During electrochemical corrosion, current flows between the anode and cathode through the body of the metal and returns through the electrolyte. If this circuit is broken, corrosion will cease.

These articles discuss methods of breaking this circuit by cleaning out corrosion cells and keeping the electrolyte from the steel by coating.

Another method is by calculating the current density needed to maintain a barrier at all areas to be protected, and imposing a direct current through the structure. Elevated water storage tanks, underground pipe lines, underground transmission lines and metal building pilings (as for the Cape Kennedy Moon Rocket Assembly building) are protected cathodically.

Cathodic protection using sacrificial anodes and impressed currents has distinct advantages. However, this method is not considered practical for cooling towers atop buildings.

Fig. 1 — Walk around inspection with clip boards to take notes.

Fig. 2 — Water leaking through louvers causes splashout and wetting of area.

Winter shutdown for spring turn on

■ Cooling tower shutdown means more than pulling switches and punching "off" buttons. Preparations must be made for winter preservation so that spring turn-on can be accomplished with minimum deterioration, effort and cost. Energy, water, and dollar conservation are today's important bywords. All of these objectives can be accomplished in varying degrees by proper inspection and maintenance during downtime.

Prior to turn-off, an inspection is in order to determine what, if any, work must be performed during turn-around period on the following systems:

— External structure
— Wet decking fill
— Water distribution
— Drift eliminators
— Mechanical equipment
— Internal structure
— Winterized cell

External Structure

A walk around (**Fig. 1**), accompanied by your assistant (if you have that luxury) together with a clipboard and pencil to make notes, is all that should be required at this time. While the cooling tower is on, observe the exterior for leaks and the security of side walls, whether they be cement asbestos, which may be insecurely fastened by rusted or broken clips; steel side plates, which may be dangerously corroded; or wood sheathing that is rotted.

Leaks concerning water or air are very costly in energy and fluid loss. Water leaks (**Fig. 2**) require extra make-up and eventually cause damage to the structure. Air leaks (**Fig. 3**) result in short circuiting and bypassing of the fill, which costs energy and does not cool water. All leaks in the structure must be filled.

Air intake louvers are important to direct flow properly into the tower and to prevent water splash out. Observe that they are in position and not broken. If replacements are to be ordered, measure their size, with particular attention to the thickness of the material, and note the amount required (**Fig. 4**).

While the tower is still operating, look into the cold water collecting basin and observe the splash pattern. Damaged, collapsed, or badly misaligned fill will be indicated by lack of water pat-

►

Fig. 3 — Air leaks through deck plates or casing causes short circuiting and loss of efficiency.

Fig. 4 — Accurate measurements must be taken for replacement of missing and broken air intake louver blades.

Fig. 5 — Droplets of drift escaping tower in spite of temporary repairs.

Fig. 6 — Inferior fiber drift eliminators, when not fully encased in stainless steel or impregnated with chlorinated rubber, have very short service life. Note corrosion on steel column.

Fig. 7 — Fill gear box when shut down for extended periods. Drain, flush, and refill with fresh oil before start up.

Fig. 8 — Combination of wood rot and steel corrosion must be corrected.

tern or excessive cascading sheets of water. Indicate on your check list areas in question which are to be further investigated when tower is turned off and dry. If areas to be rebuilt are small, redwood is easily obtainable from commercial sources. If too much of the fill has collapsed — in some cases fill slats can block off air in large areas like a closed venetian blind — the caved in fill should be taken out and replaced.

Investigate the cost of installing cellular fill against replacing all the redwood slats. While cost of the material may be greater than wood, labor is considerably less and fireproofing of the area plus increased efficiency could warrant the larger expenditure.

The cold water collecting basin splash pattern will also indicate water distribution performance. If nozzles are clogged, dry areas will be apparent, and broken arms, missing nozzles, and other imperfections in the spray or water distribution can be observed. These areas are to be noted and further checked visually when the tower is turned off.

Clogged nozzle orifices can easily be opened by removing and cleaning or by inserting probes into the nozzles to loosen debris. Broken arms to be repaired are noted and consideration can be given to replacing steel and wood distribution systems with PVC and ceramic components.

Crossflow water distribution can readily be observed when the tower is either on or off by the appearance of hot water pans and condition of metering orifices. These areas are often clogged with debris from the surroundings and scale from riser pipes. Stones, cigarette wrappers, and other foreign objects can be thrown into the tower. These are all to be removed.

Diffusion decks beneath water distribution basins should be checked for alignment, security, and placement. If wood slats are too badly broken up, they should be replaced and consideration given to cellular baffling, which breaks up solid columns of water within 4″ of top basin.

Drift Eliminators

When the tower is operating, drift eliminator units can easily be checked by observing the amount of water thrown out of the fan cylinder **(Fig. 5).** On building installations this could mean staining of bricks, etching of windows and marble work, droplets raining down on people, and invitation of a police summons. The environmental protection people are also concerned, because they feel that excessive amounts of water thrown out of cooling towers is wasteful. In industrial locations, one of the worst things that can happen to the operating engineer is that the boss's car gets wet and discolored by water treatment chemicals. He will demand and receive prompt action.

Eliminators must completely enclose the plenum area and have no deteriorated holder frames or broken blades. Eliminators must abut one against the other very tightly, because any space over 1/8″ is an invitation for water to be drawn out. Repairs can easily be effected by replacing broken wood blades and realigning eliminators. If the system is beyond repair, or where fan blades have been speeded up to obtain more capacity at a higher air velocity, consider replacing old style drift eliminators with fireproof, cellular, asbestos units which, at lower static pressure, can "wring" entrained droplets out of the airstream **(Fig. 6).**

Fig. 9 — Winterized cell illustrating bypass pipe and valve which should be maintained.

Mechanical Equipment

The greatest cost associated with cooling towers is electricity to drive fans which generate air movement through the tower. The cooling effect of the air is what makes the tower work. Anything that can be done to use every pound of air efficiently reduces cost of operation. When one considers that a 2000 ton system in a large installation (equivalent to 6000 gpm) might cost about $200,000 per year to operate, a 10% savings of $20,000 for about $1000 worth of mainnance can make you a hero in your ›ss's eyes, whether he is the factory manager, building owner, or management agent.

Of prime importance is the electrical motor. Normal maintenance and inspection, such as checking tightness and weather-protection of all connections to motor, starters, and switches, should be carried out regularly. While operating, feel motor to determine that it is not running hot due to malfunctioning wiring or bearings. Check bolts for wear and security of tiedown. With regard to V-belts, the reader would not be in an engineering position of any consequence if he need be reminded of misalignment, tension, or corrective action to be taken if misalignment is detected between sheaves of the motor and driven unit.

On large reduction gear propeller type towers, gearbox bearing failure is common. Best place to check for gear noise indicating bearing failure is at the air intake louvers. While tower motor is turned off, check for fan track and shaft rotation, together with shaft couplings for security, wear, and alignment. If so indicated, it is particularly inexpensive to replace flexible connectors, regardless if they be stainless steel discs rubber grommets. Rotate the shaft -ack and forth to determine that there is sufficient backplay in the gearbox. Examine for oil leaks, which indicate need to replace seals.

Oil level of the gearbox is important. Manufacturers recommend that if an extended shutdown period is involved, such as over the winter months when air conditioning and refrigeration is not required, that oil be filled to the top of the box to prevent condensation and rusting of the gear reducer interior. **(Fig. 7).** When this is done, it is urgent to tag the starter switch so that the cooling tower will not be activated until oil has been drained, old fluid flushed out, and refilled to the proper level with the recommended weight of oil. At this time also, check oil lines to exterior of fan cylinder to determine that they will operate properly when turn-on time arrives. If the gear transmission is filled to the top with oil, it is very important to tie the fan blades down so that they will not rotate. Windmilling of the fan or inadvertent start-up will cause internal pressure which will blow out oil seals and perhaps do additional damage.

Internal Structure

An internal inspection is necessary to ascertain structural members' condition. Sharp blows with a hammer can indicate corrosion on steel and rot on wood members. Visual inspection will also indicate condition of corrosion control coating on steel. Recheck fill, water distribution system, and drift eliminators for alignment and security plus proper configuration for providing necessary function. Also determine condition of cold water basin, whether it employs steel or wood framing. In crossflow towers check the hot water basin — steel for corrosion and integrity, wood for rot, security, and alignment **(Fig. 8).**

Winterized Cell

The winterized cell should be readied for operation. The most efficient winterization configuration is to have an auxiliary spray system with valves so that water does not travel through the fill when conditions are cold enough to permit bypassing the packing. Water returned to an absorption refrigeration system too cold will freeze the lithium bromide. Freezing temperature water is not desirable during winter. Normal differential of the cooling tower is all that is required. At times the atmosphere is cold enough that heating coils are required in the cooling tower to maintain circulating water temperature differentials. Toward this end, many operators will install an auxiliary spray unit that bypasses the fill to eliminate icing in that area and to prevent too cold a temperature occurring. Bypass valves and piping should be inspected for operation and integrity **(Fig. 9).** If water is run through the fill, icing damage could occur if air louver blade intakes are out of place or fill is improperly aligned.

Conclusion

The most important part of the cooling tower is the mechanical equipment, to which the operator must devote the most attention in order to maintain optimum cooling tower production. Downtime and associated losses can be prevented by proper operation and maintenance. Downtime not only costs money in repairs, but represents lost production. Good housekeeping and the ability to plan ahead will reduce costs and provide adequate cold water for process equipment, condensors, or turbines.

Fig. Check wood cold-water basins for deterioration, warped or split sides, open joints, and rot. Leaking basins are readily apparent during winter operation.

Starting Up Tower

Normal Operations

A. For starting when the wet bulb temperature is above 32°F and no tendency for the temperature to fall, the following should be observed:

1. Check that the water in the basin is at proper level.

2. Ascertain bypass system off (all bypass valves closed).

3. Blow-down and make-up systems to be checked for proper functioning.

4. All cell shut off valves (if supplied) fully open.

B. Start Circulating Water System

1. Activate pumps one by one to avoid any sudden shock to the tower distribution system.

2. Sufficient time must be allowed between starting each pump to balance the system flow and eliminate any blockages.

C. Starting the Fan

1. If system has been dry over one month, run water over tower at least two days before starting fan to eliminate vibrations which may occur when timbers dry and shrink.

2. Ascertain that motor instructions are followed for starting.

3. Be sure no personnel are standing in the vicinity of the fans.

4. Check to see that no foreign substance prevents free rotation of fan.

5. Check oil quantity in reducer for normal level.

6. Depending upon the load and circulating water volume, incremental savings may be realized by turning off extra fan capacity to utilize minimum number of cells.

7. If vibration switch cuts out, do not restart fan without ascertaining and correcting cause.

Safety Hints

The listed items are essential to personnel safety and are useful to longevity of the tower. All persons with cooling tower responsibility should become familiar with these notes.

A. Smoking and Open Lights - There are many parts of the cooling tower combustible. Therefore, smoking and open lights should be prohibited on or in the tower.

B. Mechanical Equipment Maintenance - Electrical lock out procedures should be employed when personnel are in contact with the mechanical equipment.

C. Vibration and Noise - Unusual vibration or noise should not be ignored. It is essential that any occurrence be investigated and remedied. Vibration switches should be wired to close fan motor off if excessive vibration occurs. An electric motor should never be reactivated until the reason for a shutdown is determined and remedied.

D. Drift Eliminators and Fill - While servicable under most normal load conditions, drift eliminator sections and fill do not constitute safe work surfaces. The condition of the drift eliminators can be determined from the access walkway in each cell and the condition of the fill and distribution system can generally be ascertained from the air inlet opening. If access is required to the drift eliminators or fill, planking should be used that will span column lines. It is extremely important that the planks be removed after maintenance work or inspection is concluded.

E. Access to Moving Parts - A hatch at the fan deck level and interior walkway is provided in each cell for inspection of fans and fan drives. Personnel entering any cell to inspect equipment should be familiar with the mechanical components and understand the potential safety hazards of moving parts. A lookout "buddy" system is essential for personnel entering tower interior.

F. Coupling Guards - If removed for any reason, the coupling guards should be reinstalled prior to activating the equipment.

G. Lubrication - Manufacturer instructions regarding lubrication should be followed strictly. It is particularly important not to over-lubricate bearings.

H. Cold Water Operation - Ice formations in the cooling tower are safety hazards and can possibly damage tower severely. Icing procedures should be activated when the ambient temperature is 40°F and falling.

I. Wood Component Moisture Content - Cooling tower wood components absorb moisture and become saturated. If operated intermittently, it is advisable to wet the tower on an occasional basis to avoid checking and splitting in the wood, particularly if weather conditions become relatively dry.

J. Water Chemistry - Evaporation of cooling tower water changes the circulating water chemistry. Changes should be monitored and an optimum program of blowdown and circulating water treatment should be implemented by a qualified chemical treatment firm. Effective control of the water chemistry will contribute to the longevity of the cooling tower components and other heat exchange equipment serviced by the circulating water.

CHAPTER 14

INSPECTION FOR PROFIT

COOLING TOWERS — How to inspect for profit

■ A well functioning cooling tower makes money by producing colder water for use in refrigeration, cooling, or process water. A 1°F colder water return from the cooling tower, under similar conditions of load and temperature, can add up to thousands of dollars saved in energy consumption or product results. In order to keep a tower functioning efficiently, all of its components must be put in optimum working order to eliminate excessive energy charges.

A 1°F increase of water temperature coming off the water tower requires an additional 3% kilowatt requirement to obtain the same refrigeration performance of the air-conditioning system according to calculations using the pressure enthalpy charts.

Therefore, if a cooling tower is returning water 3°F warmer than its design, approximately 10% more power is required. In hard cash, for a 1,500 ton compression system, at $.095 per kilowatt hour, the extra energy burden is $35.45 per hour; $354.50 per 10 hour day; $2,127.00 per week; $9,146.10 per month; *$73,168.80 per 8 months of operation.*

Obviously it behooves the operator to carefully inspect his tower to eliminate any deficiencies that prevent optimum cold water temperatures returning off the tower to the compression and condensing equipment, making note of these problems and budgeting money and time to effect repairs. A lot of upgrading can be accomplished on a 1,500 ton system to save $73,168.80 at August 1975 power rates.

Chemical plants manufacturing a salable condensate product can turn out over $1,000 a day additional materials, which will more than pay for an extensive upgrading and rebuilding of large industrial cooling towers. Power generating plants use their electricity to generate power and the degree of consumption is called the energy penalty. Again, colder water uses less energy and reduces this penalty, thereby producing more salable electricity.

There are many ways to obtain better performance from a cooling tower. The least expensive is by application of common sense, ordinary maintenance. Over the years there is "slippage," whereby equipment is reduced in efficiency in accordance to Murphy's Law, which states: "When a situation is bad and going downhill, it can do nothing but get worse." A corollary is: "By itself, equipment will not get better."

FIG. 1 — Location of cooling tower can make maintenance difficult.

The oldest and best tool available to maintenance personnel is the Mark II Eyeball. Proper inspection and adequate attention paid to discrepancies will more than pay for themselves in smoother operation. Mechanical equipment as complex as a cooling tower should be inspected with a prepared checklist so that all pertinent items are covered.

Now that the cooling season is drawing to a close, operators of air-conditioning and refrigeration systems are planning fall shut-downs and scheduling normal maintenance of the compressors, turbines, shell and tubes. The cooling tower, an important element of the entire system, because of its typical location **(Fig. 1)**, is often ignored until "five minutes" before turn-up in the spring.

Many owners and large industrial users rely on periodic inspections and reports by a professional field engineer who spends at least one to two days at the site and submits a detailed evaluation of discrepancies noticed and suggested repair programs together with possible capability of thermal upgrading.

This article discusses the procedure which parallels the methodical manner in which field inspections are made.

While there are a great variety of cooling tower types, the elements of all units are the same and a checklist can be developed covering all items as a pre-planning procedure for developing a comprehensive inspection program.

There are three major divisions of a cooling tower: *exterior structure, mechanical equipment,* and *interior structure.* We will investigate each division as a unit, marking down discrepancies as they appear.

1. **EXTERIOR STRUCTURE: (Fig. 2)** This inspection is the visual walk around examining the following components:

   A. Louvers: Determine the number of missing louvers, measure accurately the existing louver size, length, width and measure thickness very carefully since 1/16″ difference can prevent the new louver blades from fitting in the existing support slots. Out-of position louvers should be checked for repositioning. If the louvers are wood, look for splits or breakage and check for wood rot and decay. Asbestos cement or fiberglass reinforced polyester louvers are to be examined for breaks, cracks, and corner damage which prevents proper alignment. Examine louver supports. If made of wood, for rot and decay; if galvanized steel, observe condition of protective coating and make notes of broken and rotted supports. See that attaching hardware is adequate and properly supporting the louver posts. Examine louver supports for ice damage.

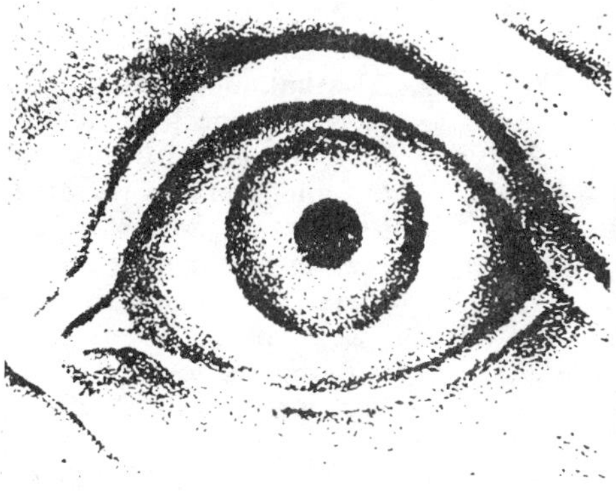

The 'Mark II Eyeball' is the power engineer's best maintenance tool,

FIG. 2 — Walk-around inspection includes notation of missing/broken louver blades.

FIG. 4 — Installation of fiberglass siding to replace rotted wood.

B. Casing: On corrugated or flat cement asbestos board, transite, or fiberglass reinforced polyester casing, look for cracks, leaks, condition of attaching hardware and other physical damage. Corner rolls should be observed for integrity and security of attachment. On wood casing towers look for loose boards, evidence of wood decay **(Fig. 3),** leaks and wood rot. Wood for casing of towers is a rather old technique and usually consists of a double layer. This should be checked from the interior to determine if wood rot dictates entire replacement with modern fiberglass sheathing **(Fig. 4).**

C. Access Door: Check for missing doors, broken or loose hinges or latches, general condition of doors in operation **(Fig. 5),** and any evidence of physical damage. Measure door accurately if replacement is necessary.

D. Distribution System: Observe flange connectors at top of riser for security and gasketing and make notes of effectiveness of corrosion control coating on supply lines and risers. On counter-flow towers check where header enters casing **(Fig. 6)** to determine if it is properly caulked and no apparent leaks exist. On cross-flow towers look for deterioration in distribution basins, splash guards, and associated piping. With water trough configuration, check boards for warpage or splitting on basin sides and leakage due to gaps in trough.

FIG. 3 — Wood rot of exterior vertical sides necessitates replacement.

FIG. 5 — Check operation of access door hinges.

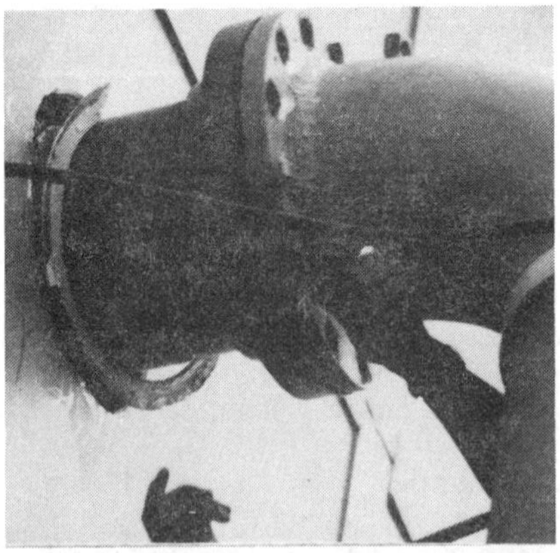

FIG. 6 — Piping entering transite wall must be water-tight.

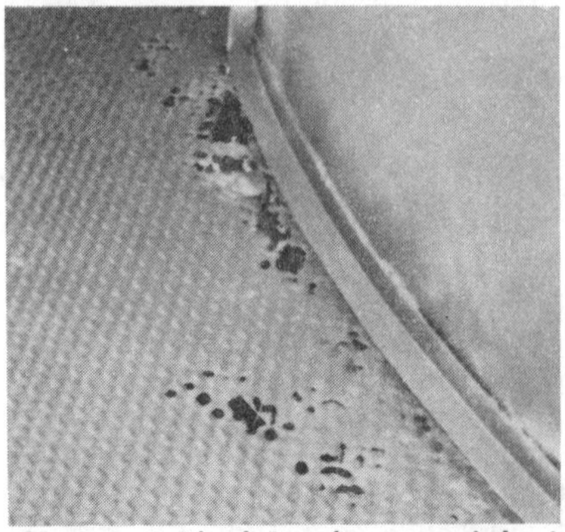

FIG. 8 — Besides being dangerous, holes in deck plates short-circuit air movement, lowering performance.

FIG. 7 — Observe cantilever supports before climbing stairways.

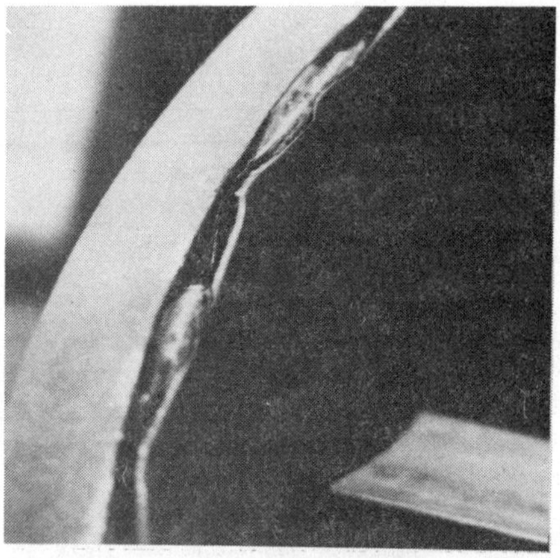

FIG. 9 — Corrosion on fan cylinder can move steel into path of rotating fan blade.

FIG. 10 — Nozzles should be checked for proper water distribution pattern.

FIG. 11 — Iron rot occurs at interface of wood and steel.

FIG. 12 — Mis-aligned cross-flow drift eliminator blades create drift problem.

FIG. 13 — Ice damage at air inlet louvers will break wood fill slats.

FIG. 14 — Interior rotted wood supports must be changed.

E. **Drain Boards:** Look for evidence of water being diverted outside drain to tower exterior. Check for displaced or damaged drain boards. If drain boards are of metal, observe for corrosion and rot. Irrespective of materials used, look for missing boards and condition of fasteners.

F. **Stairway:** Note deteriorated or loose hand rails and kneerails. Check stringers and stair treads for missing components, breakage and cracks. Examine structure for looseness, evidence of rot, and condition of fastening devices. Steel stairways should be checked for evidence of corrosion or acid attack. All fasteners should be examined for security (in most installations, it is prudent to "shake" the stairway before using). Special attention should be paid to condition and security of upper braces of cantilevered platform supports **(Fig. 7).**

G. **Ladder & Walkways:** Before climbing any ladder, check for security of fasteners by vigorously shaking ladder rails, and then observe that all the rungs are in position. Note missing or deteriorated rungs. Check treads on walkway for security, rot, or corrosion. Check stringers and rails for corrosion. In order to comply with O.S.H.A. safety requirements, any ladder over 20′ must have a cage or safety device to protect the climber. Further, a cage above 20′ must have welded platforms at 20′ interval or be protected by a safety device. Check these appliances for security and evidences of corrosion, rot or weakness. An old surviving "ladder climber trick" is to grasp ladder rails rather than rungs since rungs could be rotted and weak.

H. **Fan Deck:** Examine fan deck for decay, missing or broken members, and gaps between boards which can cause short circuiting of air circulation. On steel decks check for corrosion, holes **(Fig. 8),** and condition of corrosion control coating. Fan deck supports should be examined for decay, loose or broken members, and condition of fastening devices.

I. **Fan Cylinders:** Check that fan cylinders are securely anchored to the fan deck supports. Observe for looseness of fan cylinder and condition of fastening devices. Look for damage, missing parts of the structure, and deterioration of components. Observe wood rot, or where applicable, corrosion of steel sides, rings **(Fig. 9),** or support strap banding. Inspect for proper tip clearance between fan blade and interior of the cylinder walls. O.S.H.A. safety requirements dictate a protective fan guard on cylinders less than 5′ high. If guard is installed, check condition of components for corrosion, sagging, and missing parts.

2. INTERIOR STRUCTURE

A. **Distribution System:** Inspect metal piping headers for decay, rust or acid attack. Check flange connections for tightness and condition of fasteners and gaskets. See that all branch arms are secure; note any corrosion on flange connectors. Check for missing nozzles and, if practical, observe spray pattern to see if nozzles are in

FIG. 16 — Inspection spotted broken bolts (arrows) which prevented very expensive fan wreck.

FIG. 17 — Excessive play and noisy bearings necessitated field change of gearbox bearings.

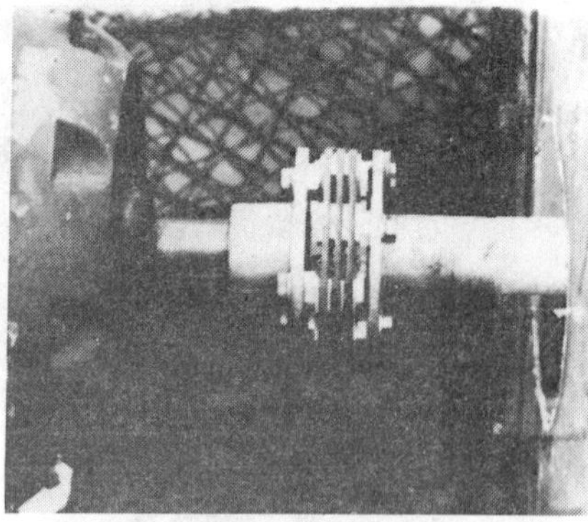

FIG. 18 — Inspect connectors of flexible couplings.

good operating order **(Fig. 10)** and not clogged. For cross-flow water distribution, see that nozzles are cleaned and operating and, if nozzles are not used, check that all distribution holes are clear and not clogged. Note condition of redistribution section beneath hot water basin for necessary repairs or replacements. If piping is fabricated from wood, check condition of retaining bands, look for warpage or split boards, and check for excessive leakage. Water trough distribution should be observed for condition of wood, fasteners, excessive gaps between sides and bottom of trough, and alignment of splash cups.

B. Mechanical Equipment Supports: Inspect the mechanical equipment supporting steel for evidence of corrosion and condition of the fasteners, straps, or hangers. Look at wood structural members in contact with the steel supports to ensure soundness or evidence of weakness which could result in loss of structural strength. If springs or rubber vibration absorption pads have been installed check conditions of springs, mounting and adjusting bolts, corrosion of ferrous members, and condition of rubber pads. Check for "iron rot" where steel and wood are in contact **(Fig. 11).**

C. Fire Detection: Test valves and operating condition of wet or dry sprinkler system. Check for evidence of corrosion on pipes and connectors. See that activating devices are in apparent good order. If a thermocouple wiring system is used as an alarm indicator, check for broken wiring.

D. Eliminators: Check drift eliminators for broken, displaced or deteriorated blades **(Fig. 12)** for gaps or misalignment which will permit excessive drift. Check supports for breakage or deterioration. Look for clogging due to algae, debris, or slime. Split or expanded cross-flow fiberboard or plywood vertical support members should be examined for possible replaeement.

E. Fill: Inspect tower fill for any breakage, deterioration, misplaced or missing splash bars. Look for damage to splash bars supports and fill supports **(Fig. 13).** If sections of fill have collapsed look for damage from icing. Obtain design conditions of water and operating temperatures so that thermal calculations can later be made to determine degree of upgrading capability available by changing fill to a more efficient design. Fireproof asbestos fill should be considered as an alternate to expensive automatic deluge system since piping and controls of a fire alarm system need continual checking and maintenance, while asbestos fill, by the nature of its fireproof construction, does its fire pro- ►

FIG. 15 — Corroded steel should be sandblasted. Note cell dividing basin flanges. Dividing walls should be added from bottom of fill to basin to prevent air short circuiting on limited cell operation.

FIG. 19 — Inexpensively installed shaft guard (arrows) can prevent expensive fan wreck if coupling breaks.

tective job without further attention.

F. Interior Structural Supports: Columns, girts, and diagonal wood members can be tested for soundness by striking with a hammer. A high pitch sharp sound indicates good wood, while a dull low key sound indicates softness of wood. Wood rot will soften wood **(Fig. 14)** and if an area indicates possible rot, probe with a screwdriver to further determine condition of the wood. Check carefully around metal fasteners for "iron rot" on wood in contact with metal. Check steel internals for corrosion and deterioration due to rusting. Check bolt condition and tightness on both wood and steel interiors.

G. Partitions: Check bolts and nuts in partition for looseness and corrosion. Inspect for deteriorated or loose partition boards. Note if partitions go from basin to roof deck dividing individual cells so as to prevent windmilling of idle fans or short-circuiting of air between cells. See that wind walls parallel to air intake louvers are in position, that boards or transite members are securely fastened. Check supports, either wood or steel, for condition of rot or corrosion and condition of the protective coating.

H. Doors: Make notation of missing doors, condition of hinges and fasteners, and look for corrosion, rot, or deterioration of door material. If cell partition does not have doors at basin level, so indicate on report.

I. Cold Water Basin: For wood basins check deterioration, warped or split basin sides, open joints, and soundness of wood. Inspect steel basin for corrosion, holes, and condition of protective coating **(Fig. 15)**. Concrete basins should be inspected for cracks, breaking of concrete, spalling, open joints or acid attacks. All sumps should be observed for accumulation of debris, condition of screens, anti-turbulator plates, and freedom of operation of drain valves. If basin is one continuous unit, partitions between cells together with necessary sumps and piping should be recommended to add flexibility for utilization of equipment by operating only the number of cells required for the heat rejection load, thereby conserving power.

3. MECHANICAL EQUIPMENT

A. Fans: Check fan hubs for corrosion. Inspect hub covers for corrosion and condition of attaching hardware. Inspect blade clamping arrangement for tightness and corrosion **(Fig. 16)**. Look at fan blades carefully for corrosion and erosion. Check fan blade pitch with bubble protractor for uniformity; check blades for build-up of solids which would change blade moment weight or air foil characteristic, thereby causing excessive vibration.

B. Gear Boxes: Check gear reducers for proper oil level, inspect oil for moisture and sludge. Rotate input shaft by hand back and forth against gear tooth contact to feel for back-lash. Check input pinion shaft bearing for wear by attempting to move shaft radially. Look at fan shaft bearings for excessive end play by applying force up and down on a fan blade tip and note the movement of output shaft **(Fig. 17)**. *A running clearance is built into some output shafts which should not be confused with excessive end play.*

C. Power Transmission: Inspect drive shaft keys and set screws; check assembly hardware for tightness. Examine drive shaft couplings for corrosion, wear, or missing elements. Look at exterior of drive shaft visually for corrosion and check interior by lightly tapping drive shaft for sound along its length for dead spots which could indicate internal corrosion. Carefully observe flexible connectors of both ends of shaft **(Fig. 18)**. A shaft guard is imperative to prevent fan wrecks. The weakest length in the power transmission train is the coupling connectors. Invariably when a coupling breaks the rotating shaft swings upward and wrecks fan blades plus possibly doing damage to gear box and motor. This can be prevented by the installation of simple, inexpensive shaft guards **(Fig. 19)** to absorb the upward thrust of the rotating broken shaft. Belt driven units should be observed for alignment, tension, and condition of wear of belts.

D. Motors: Look for evidence of motor heating; observe proper lubrication of the bearings. Excessive grease or oil applied to ball bearing motors can damage the unit by forcing grease into windings and causing deterioration of insulation. Look for deposits of dirt or dust at motor air intakes and check TEFC motors for conditions of air passages and fans. Drain moisture plug is supplied with motor; see if it is operational. Measure operating loads with amp probe and voltmeter as compared with name plate data; recommend pitching of fan blades up or down to compensate. Examine mounting bolts and attachments for security and corrosion.

All of the above items might seem excessively complicated and a lot of work. However, if a thorough investigation and description is to be made, it must cover all elements of the cooling tower. To facilitate the actual inspection, a check list can be compiled (see sample), which will greatly speed up the job and ensure that all aspects of the inspection were accomplished. This will also serve as a reference to writing and submitting your report.

Inspection and Maintenance Checklist

A. Introduction

1. Planned routine periodic inspection and maintenance procedures are very necessary to prolong the life of the cooling tower. Mechanical equipment requires inspection and lubrication. A prepared checklist is essential for proper maintenance.

2. All lubrication schedules should be maintained and documented.

3. Unusual noises should be sourced and vibrations detected early before damage occurs.

4. Heat transfer can be kept at optimum levels by maintaining uniform water distribution through clean nozzles, clean and unbroken water distribution piping, properly operating valves, and uniform fill materials. Pumps, motors, and related equipment should also be inspected.

5. Safety is paramount. Inspection and maintenance operations are to be carried out so as to not cause injury to personnel.

6. A quick visual inspection tour should be made of the tower once each shift to assure that no problems have developed. A report to the plant control room operator or an entry in the plant log should be made at this time.

B. Start-Up After Down Time

1. Gear Reducer

   a. Check that lubrication is of recommended type and proper level and that all oil lines, valves, and fittings are tight.

   b. If oil filter is supplied, it should be air bled and bypass valves properly closed.

   c. Grease fittings, if supplied, are to be checked for type and amount.

   d. All mounting bolts are to be tightened to torque specifications.

2. Electric Motor

   a. Mounting bolts should be properly torqued and in required alignment.

   b. Adequate grease present in the motor bearings.

   c. Electric connections are tight.

   d. Correct rotation of the motor shaft should be checked.

3. Mechanical Coupling

   a. The motor end of coupling should be checked for alignment and inspected for cracks, warped plates, and tight connections.

   b. See that pillow block bearings are lubricated with roller bearing grease suitable for 1800 RPM speed and high moisture exposure.

   c. Check coupling to gear box end for alignment, cracks, warped plates, and tight connections.

4. Fan Assembly

   a. Fan blades checked for proper pitch and uniformity of pitch angle.

   b. Fan tip to the wall clearance should be 1¼ to 2 inches.

   c. Hub bolts are to be properly torqued.

   d. Look for blade cracks or physical damage.

   e. Check direction of rotation for proper air movement.

5. Vibration Safety Cut-out Switches

   a. Check switches for proper operation and cut-off limits according to specifications.

   b. See that components are free to move at the proper level of sensitivity.

6. Vibration Signature Analysis

   a. Mechanical equipment vibration monitoring and analysis should be performed at start-up.

7. Gate Valves

   a. If motor driven, check to determine proper alignment of components, tighten, waterproof electrical connections, and proper relay action.

b. The units should be timed to determine opening and closing time in accordance with manufacturer's instructions.

c. Excessive time would indicate a bind or malfunction which must be corrected.

d. Check proper motor rotation.

8. <u>Thermal Transfer</u>

a. Fill material should be inspected for integrity, uniformity, and any unusual wear.

b. Drift eliminators examined for broken blades or malfunctions that are correctable.

c. Nozzles can be inspected for malfunction by noting water pattern in cold water basin.

d. Air intake louvers visually inspected to assure proper function.

9. <u>Structural Members</u>

a. Structural components should be periodically observed to determine their overall condition.

b. Connectors should be observed for tightness and if fabricated from ferrous material, should be monitored for corrosion.

c. Hand rails should be in place and firm; ladders, stairs, and walkways must also be secure.

d. Hatches and access panels should be kept closed and well maintained.

e. Drift eliminators will not support workmen under normal conditions and should be planked. Planks to be removed after maintenance and inspection is completed.

f. Beware of mechanical components in motion (i.e., fans, shafts, or motors) and observe plant safety lockout procedures

COOLING TOWER INSPECTION CHECK LIST

RBA Job No.

Owner	Date Inspected
Plant	Inspected By
Location	Tower Manufacturer
Owner Designation	Installed 19
Water Treatment Used	Model No.

Design Conditions ______ GPM ______ HW ______ CW ______ WB

Condition: 1 - Good; 2 - Repair; 3 - Replace	*1*	*2*	*3*
EXTERIOR STRUCTURE:			
1. Endwall Casing & Access Doors			
2. Louvers ( )			
3. Drain Boards			
4. Stairway			
5. Fan Deck			
6. Fan Deck Supports			
7. Handrails			
8. Ladders & Walkways			
9. Distribution System			
Headers (Type)			
Distribution Basin			
Water Level			
Flow Control Valves (Size)			
Nozzles (Size)			
Water Distribution			
10. Spray System & Spray Nozzles			
11. Fan Cylinders (Type)			
INTERIOR STRUCTURE:			
12. Fill (Type)			
13. Columns			
14. Girts			
15. Diagonals			
16. Partitions & Doors			
17. Eliminators (Type)			
18. Walkway			
19. Cold Water Basin (Type)			
Water Depth			
20. Mech. Equip. Support (Type)			

Condition: 1 - Good; 2 - Repair; 3 - Replace	*1*	*2*	*3*
MECHANICAL EQUIPMENT			
21. Drive Shafts (Type)			
22. Speed Reducer			
Series ______ Ratio			
Oil Level			
Oil Seals			
Vent			
Back Lash			
Pinion Shaft Play			
Fan Shaft End Play			
Last Oil Change (Date)			
Oil Used			
23. Fans			
Dia. ______ Type			
Hub			
Blades			
Hub Cover			
Tip Clearance			
No Vibration ______ Vibration			
Additional Components (If installed on tower)			
Fan Guards			
Oil Gauge & Drain Lines			
Vibration Limit Switches			
Other:			
24. Motor: Mfr.			

Name Plate ______ HP ______ RMP

Phase ______ Cycle ______ Volts ______

Amperes ______ Frame ______

REPLACEMENT PARTS REQUIRED:

QUANTITY	DESCRIPTION	ORDER FROM	DATE REQ'D

MAINTENANCE WORK REQUIRED:

DESCRIPTION	REQ'D COMPLETION

(Use back of this sheet for additional requirements or notes.)

CHAPTER 15

FAN WRECK PREVENTION

FAN WRECK PREVENTION

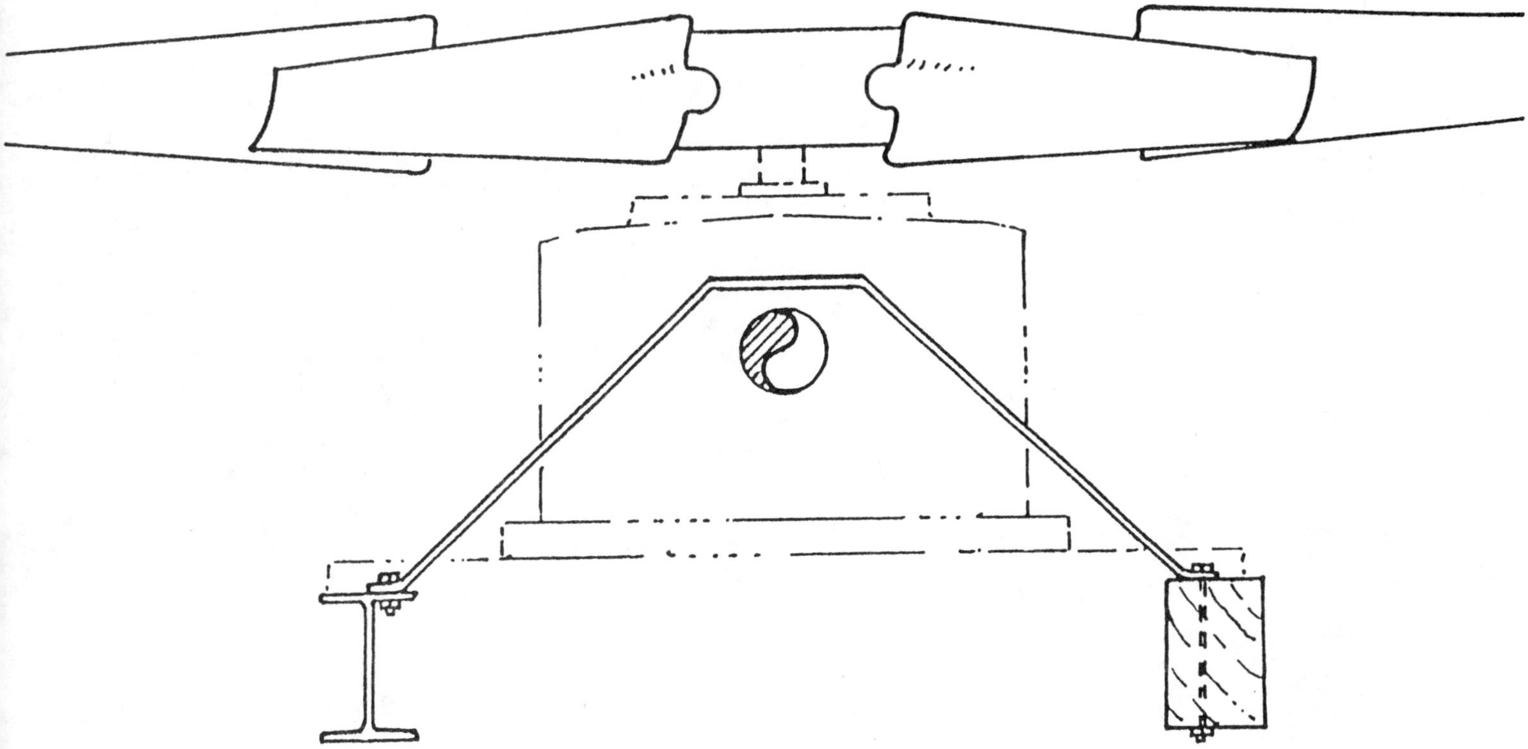

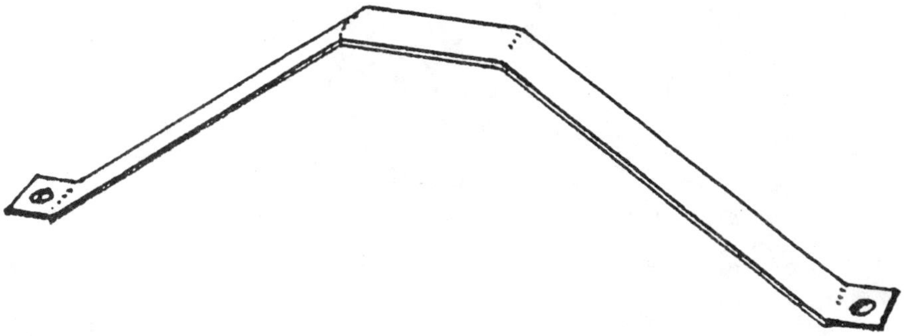

The cost of a fan wreck is measured by more than the replacement charges. What is the cost of lost production? Higher pressure heads? Lower operating levels?

Installation of a simple fan shaft guard by in-house maintenance personnel can prevent very expensive and time consuming fan wrecks.

CHAPTER 16

OPERATING CONCEPTS

Cooling tower operators are a remarkable breed of engineer. In all too many cases, they are shown the equipment, given the manufacturer's "boiler plate" sales brochures, and told to operate the machinery. While it sometimes amounts to on-the-job training, it usually turns out well due to the resourcefulness of the operator and his application of common sense and past experiences.

Cooling towers are often underestimated as mechanical equipment with vast energy-savings potential. In the past, the tower was considered on the waste heat side of the process and while it saved water compared to wasting water, it was not considered a product or revenue creating device. However, in today's energy climate, it is recognized that a fully functioning tower will use excessive energy and in chemical process plants the cooling tower is, in all too many cases, the limiting factor of production.[1]

To guide the beginner (and to pass on useful information and hints to the old pros) the following data has been compiled.

Design Conditions and "More Water" In Hot Weather

Cooling towers are designed to cool a stated volume of circulating water, entering at a specific operating temperature, leaving at a colder level (ΔT cooling range), at a given ambient wet bulb temperature. **(Fig. 1.)**

Cooling tower operation does not depend upon the dry bulb temperature. The ability of a cooling tower to cool water is a measure of how close the tower can bring the water temperature to the wet bulb temperature of the surrounding air. The lower the wet bulb temperature, which indicates either cool air, low humidity, or a combination of the two, the lower the tower can cool the water. It is important to remember that no cooling tower can ever cool water below the wet bulb temperature of the incoming air. In actual practice, the final water temperature will always be about 5° to 7° above the wet bulb temperature, depending upon design conditions. The wet bulb temperature selected in designing cooling towers is usually close to the average maximum wet bulb for the summer months at a given location.[2]

If an installation is desired to cool 10,000 gallons per minute (GPM) of circulating water from 120°F hot water (HWT) to 85°F cold water (CWT) at a 75°F wet bulb (WBT) i.e., a 35°F Range (HWT-CWT) at a 10°F approach to the wet bulb, which is defined as the cold water temperature less the ambient wet bulb temperature (CWT-WBT), it will not cool 11,000 GPM to a 10°F approach under these conditions. However, it is obvious that the tower will easily cool 9,000 GPM to design and lower.

The above examples apply to large multi-cell cooling tower complexes. However, the 10,000 GPM usually is spread into four cells of 2,500 GPM each and the procedures presented for 4 cells would apply equally to 1 cell of the same or smaller magnitude. **(Fig. 2.)**

Therefore, why, in the middle of summer, with a high wet bulb, the process department, when temperatures are rising inside the plant, keeps calling for more water! More pumps on! Flood the tower! "We need more water!" Translated: We really require COLDER water.

It takes a cool cooling tower operator made of stern stuff to ignore this plea and deliberately shut down pumps to reduce the water flow over the tower. Admittedly, an increased velocity of water through the shell and tubes will have a small cooling effect, but not as much as that provided by COLD water.

Reduced volume of water flow, below design capacity, will return colder water which will then satisfy the operation. Do not inform the process department — they won't believe you. However, if you understand what design conditions allude to, then the above procedure will be apparent.

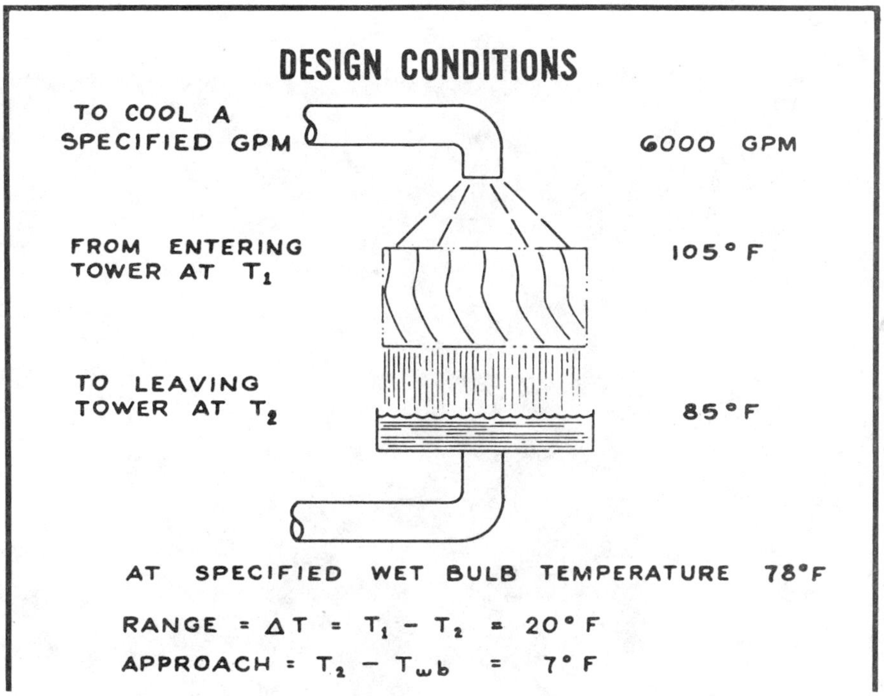

FIG. 1 — Design conditions describe the duty and work expected from a cooling tower. They consist of five variable input factors: water volume, temperature in, temperature out, and wet bulb temperature consisting of relative humidity and ambient temperature.

FIG. 2— On a cell-to-cell basis, thermal, hydraulic and aerodynamic calculations are the same for a small, one cell installation (right) as for a large, multi-cell installation (left).

In Cold Weather

Do you operate under icing conditions? **(Fig. 3.)** Be prepared when the temperature is dropping through 40°F to alert the crew for ice formation.

What is the best way to minimize icing? In a multi-cell tower, should you spread the available water over all of the cells for better cooling? Or when operating a one cell tower, uniformly pass the reduced flow over the entire cell? WRONG!

Valve off as many units as necessary to maintain full design water flow to the remaining cells, or utilize only a portion of the single cell. Small volumes of water tend to freeze up rapidly as they flow through the tower. This is especially true of those valves that are almost closed but not seated properly, permitting a trickle of water through. Part of your maintenance check list, performed during good weather, is to insure proper operation of all balancing and control valves.

When ice does form, especially at the bottom of the air intake louvre areas, an accepted technique is to operate the fans in reverse about 5 to 10 minutes every hour, according to conditions, to melt the ice buildup.

The automatic variable pitch (AVP) fan, **Fig. 4,** utilizes heat sensors in the basin which informs the pneumatic pitch controller to set the blades to optimum angle for current thermal demand load thereby utilizing less energy.

Since conventional fans and motors

FIG. 3 — Icing, if not properly handled and controlled, can cause extensive and costly damage.

are set to provide 100% maximum air delivery at all times, they are energy wasteful 75% of the time when less than full heat load is on the tower. The AVP fan manufacturers claim that the average retrofit installation will have an approximate 1-1/2 year cost payback, with an operational profit thereafter. Further, in de-icing situations, the AVP installation can be automatically programmed to reverse pitch at predetermined intervals for a measured period of time eliminating error when manual operations are relied on.

FIG. 4 — Automatic Variable Pitch (AVP) fans, besides being energy savers by sensing heat load and changing pitch, can also be automatically cycled in reverse pitch to melt ice build-ups.

Range Vs. Approach

Operator A states that his tower is cooling 10,000 GPM through a 15°F range. Operator B has a better tower because it cools 10,000 GPM through a 22°F range. Not necessarily so!

One important factor was not reported. Do you want a cooling tower with a 50°F range? Certainly, we can build it to cool the entering 150°F water to 100°F. A cooling tower does not generate nor destroy heat, it merely moves it from point A to point B. Further, any cooling tower will reject all of the heat load put on it.

A cooling tower cannot be sized on a number of BTUs per hour it will dissipate. Any cooling tower, regardless of it's size, will dissipate all the BTU's sent to it. Unlike a pump, which will only move a certain quantity of water against a definite head, a cooling tower will continue to dissipate BTU's to the atmosphere as long as they are added to the circulating water. If enough BTU's are added to the circulating water, the cooling tower will dissipate the steam coming to the tower.[3]

The important missing ingredient not reported in all three examples above is the wet bulb temperature (WBT). The difference between the ambient WBT and cold water temperature (CWT) of the tower is the approach.

The thermal capacity of the cooling tower is not the amount of heat rejected, rather it is the level at which the heat is rejected. Therefore, the measure of the performance of the cooling tower is the resultant cold water temperature relationship to the Design wet bulb or more specifically, the approach, CWT-WBT, under given conditions. Cold water temperature is the primary dependent variable, and vividly indicates cooling tower capability. For example, a cooling tower might be designed for cooling water from 110°F to 85°F at a 78°F wet bulb. This gives us the design range of 25°F at an approach of 7°F. If the cooling tower in actual operation is at these temperatures, then it is performing. However, it could be operating at the conditions of 120°F entering hot water to 95°F cold water at a 78°F wet bulb. In this instance the cooling range would be the same 25°F so it could be said that the cooling tower was cooling the water 25°F in strict accordance with design. Not true, however, the approach would have opened up from 7°F to 17°F and this cooling tower would be operating at about 50% of the original design performance.[4]

Certified testing a tower by the Cooling Tower Institute (CTI) provides an accurate parameter of its performance and indicates a percentage of capability of design which then redefines the actual operating level of the unit.[5]

It is therefore obvious that the cost of erecting or retrofitting a tower to increase the performance depends upon where the approach is. To reduce the approach from 15°F to 14°F would cost approximately 7% more while reducing the approach to the wet bulb from 17°F down to 7°F would require 50% more tower in size and cost.

To re-emphasize the importance of the approach, the thermal level of heat rejection of the cold water in relationship to the wet bulb is the criteria of the level of operation of a cooling tower.

Murphy's Law and Maintenance

1. If there is a possibility of several things going wrong, the one that will cause the most damage will be the one to go wrong.
2. Left to themselves things tend to go from bad to worse.
3. Once a job is fouled up, anything done to improve it only makes it worse.
4. Any device requiring service or adjustment will be least accessible.

Regardless of the wealth or poverty of any given maintenance department, irrespective of the sophisticated tools in an individual's tool box, every maintenance man has a fantastic tool with which to ply his trade. It is called the Mark II Eyeball. **(Fig. 5.)**

Develop a maintenance check list and use it! There is no substitute for a good checklist **(Fig. 6.)** operating in conjunction with the Mark II Eyeball. This, of course, applies to maintenance on all types of equipment, not only cooling towers.

Visual inspection can detect weak or broken couplings. Fan blade cracks which can cause expensive fan wrecks and lost production due to down time can be visually spotted. Misaligned fill and drift eliminators result in deficient cooling levels. Rotted or weakened columns, beams and braces, can be pinpointed before they become critical. All of these and more items can be programmed into a good maintenance inspection checklist.

Power Consideration

An increase in horsepower or increase in fan pitch angle, let us assume an arbitrary figure of 20%, will not increase the capacity of the tower or air volume moved by the same amount of 20%. Fan laws dictate that the improvement will be in proportion to the cube root of the power increase. That is, a 20% increase in horsepower from 100HP to 120HP will yield a 6.3% increased capacity, not an additional 20% of air.[6]

$$\left(\frac{120\ \text{HP}}{100\ \text{HP}}\right)^{1/3} =$$

$$\frac{4.933}{4.642} \times 100 = 106.3 = 6.3\%$$

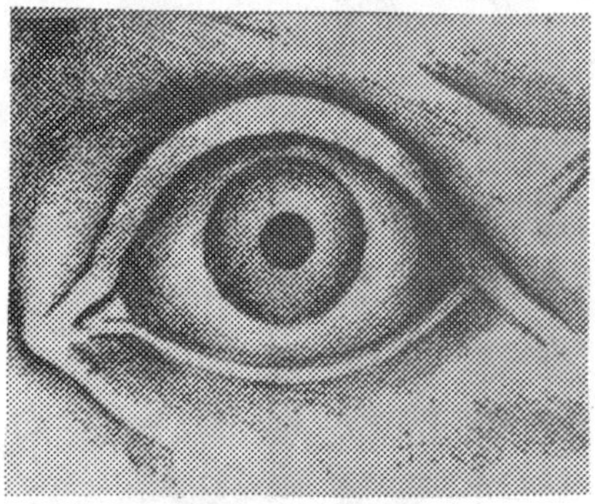

FIG. 5 — The best and most useful item in a maintenance department is the Mark II Eyeball. Used properly, this tool can be a tremendous money saver.

Evaporation Rate

Water loss due to evaporation equals 0.1% of the circulating volume per 1 0° Delta T range. This is a function of the evaporative cooling release of the latent heat of vaporization and will be quite constant for all cooling towers of the same design conditions regardless of the varied claims of manufacturers.

In the direct contact of air and water, mass transfer of heat occurs, normally from the water stream to the air stream. In normal cooling tower operation the objective is the transfer of heat from the circulating water to the air stream. Since the air stream dry bulb temperature may decrease in some cases and increase in others, it is evident that, in addition to water heat load, the entering air wet bulb and dry bulb temperatures and the air mass flow rate also influences the evaporation loss.

Driving Force and Humidity

While the power source rotating the fan is commonly called the "driver," be it an electric motor or steam turbine, the prime mover for cooling is referred to as the enthalpy driving force. At low air temperatures, about below 73°F dry bulb temperature, the moisture content of the air has very little effect on enthalpy at constant wet bulb temperature. At relatively high air temperatures, however, the effect of humidity is sufficient to cause a noticeable change. The enthalpy of very dry air will be enough lower than that of very humid air to yield a significantly higher driving force.

For a given wet bulb temperature, humid air is more dense than dry air. At equal mass flow rates the velocity through a cooling tower, and therefore resistance to flow, will accordingly be lower for the humid air.

Cooling towers are much more than those "boxes in the back that cool the water." In today's climate of ever-expanding energy costs, a knowledgeable cooling tower operator, by producing colder water from his equipment, can conserve a considerable amount of energy and earn extra profits for his company by producing additional product due to the lower temperatures in the process equipment.

For an up-to-date comprehensive evaluation, in addition to periodic inplant maintenance inspections, it is recommended that a professional cooling tower firm be utilized to look the facility over and submit a report on the modernization and energy conservation potential available by applying modern state of the art technique.

Evaporative condensation, the main principle of cooling tower operation, creates a loss of water volume in the form of vapor discharge.

References

1. "Cooling Tower Operations, J. Willa, Cooling Tower Institute Proceedings, January, 1980.
2. "Engineering Weather Data," Joint Publication of the Department of Army, Navy, and Air Force.
3. "Cooling Tower Sizing," R.W. Maze, HYDROCARBON PROCESSING, 1967.
4. "Cooling Towers, Energy Conservation Workhorses," R. Burger, ASHRAE JOURNAL, April, 1980.
5. "Acceptance Test Code For Water Cooling Towers," Cooling Tower Institute, ATC-105, February, 1975.
6. "Cooling Tower Technology Textbook," Chapter 8, Thermal Calculations, Robert Burger, 1979.

Getting rid of asbestos

During removal from a cooling tower, friable-asbestos components can flake, creating a toxic hazard. Correct procedures will avoid this problem and assure compliance with regulations.

Robert Burger, Burger Associates, Inc.

It is well known that asbestos is carcinogenic. In many cases, it should be removed from process equipment and be replaced with a safe material. Just how great is the danger from asbestos and how does it attack the body?

There is a constant life-and-death battle being waged inside the human body. The photo at right shows a 4,000× photomicrograph of a macrophage white-blood-corpuscle reaching out to envelop a protein disease-microorganism (*Eschericia coli*) in the body [*1*]. The macrophage patrols the body in pursuit of pollutants, bacteria, parasites and other disease agents. This class of white cells creeps along the chemical trail of, for example, *E. coli.*

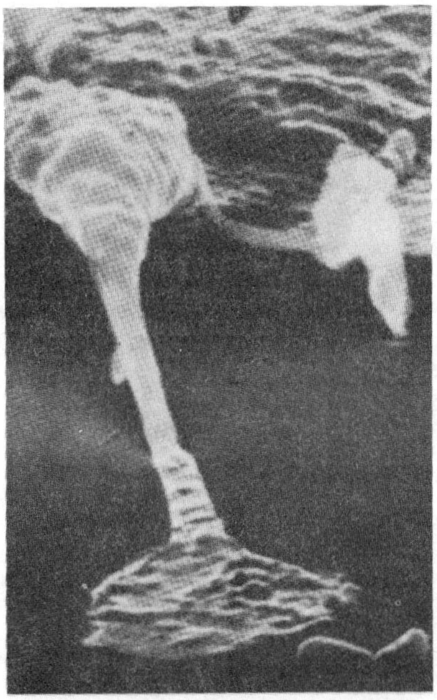

When a bacterium is encountered, the macrophage springs into action, waving its tendrils and stepping up the production of digestive enzymes. The bacterium is snared, drawn inside the cell, and destroyed by the enzymes. Having done its job, the macrophage begins searching for another invader.

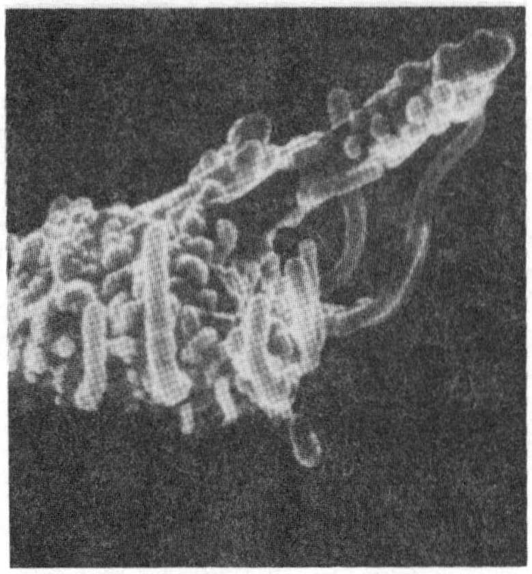

Asbestos

The above photo is a 20,000× photomicrograph of a macrophage futilely entrapping an asbestos fiber [*1*]. Although macrophages can engulf fine mineral fibers in the lungs, they cannot digest asbestos. The white cells collapse and die in the attempt. This releases the fibers back into the lungs. Another fighter "sees" the debris and, in turn, dies in trying to digest it.

This buildup is the scarring of lung tissue that can lead to asbestosis, lung cancer and mesothelioma (a tumor of the pleura). Thus, asbestos has been subjected to heavy regulations.

Cooling-tower hazards

In cooling towers, asbestos is usually found as cement asbestos, and in drift eliminators and wet-decking-fill. Many cooling towers have drift eliminators and fill made of asbes-

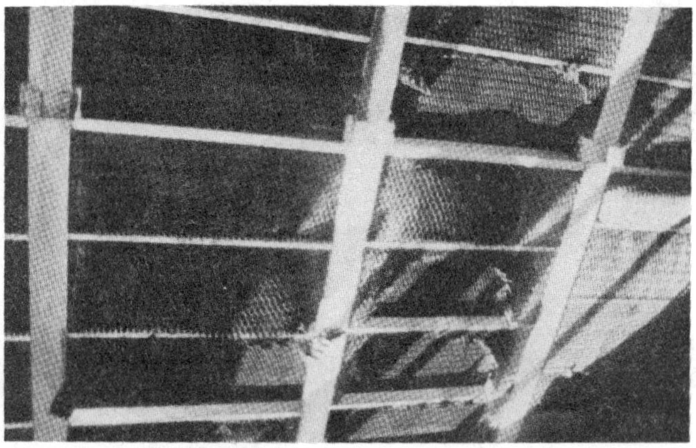

tos paper, plus siding and air-intake louvers of cement-asbestos board.

The U.S. Environmental Protection Agency (EPA) classifies asbestos as friable and nonfriable. The friable material is the hazard — it "can be crumbled under hand pressure" — and therefore is likely to emit fibers into the atmosphere.

While operating in the tower, the paper-based mist eliminators and fill, and the cement asbestos, are classified as nonfriable. Usually, the paper is impregnated with melamine, which seals-in the fibers for normal cooling-tower usage.

However, after years of service, the fibers leach out. When components are taken out of the tower to be replaced, the asbestos is then classified as friable (the above photo). At that time, the owner and contractor become responsible under mandated rules and regulations.

(When installing new drift eliminators, care should be taken with design calculations. The velocity through the devices should not exceed the resistance of the new units, consistent with low static-pressure-drop.)

Asbestos regulations

Of course, for cooling towers, the exposure to asbestos is close to a minimal level, but all regulatory agencies write asbestos regulations for the worst case. These regulations are quite detailed. For example, the National Emissions Standards for Hazardous Air Pollutants (NESHAP), as given in the *Federal Register* [*2*], set forth criteria and standards for demolition and renovation, which apply to owner and operator.

NESAHP, regulations of the U.S. Environmental Proctection Agency (EPA), state that:

If the amount of friable asbestos material in the facility being demolished or renovated is at least 80 linear m (260 linear ft) on pipes or at least 15 m^2 (160 ft^2) on other facility components, then all parts of the regulations apply.

Prior and during removal, the asbestos must be wetted, regardless of its footage. Other requirements: proper packaging, air cleaning and disposal of the material at an approved site.

When removing asbestos, NESHAP's detailed precautions should be followed, rather than not following them and monitoring exposure levels instead. With monitoring, the personal exposure limitations (PEL) of 0.2 fibers/in.^{3} action-level must not be exceeded.

It can take hours to days for a laboratory to report on the PEL level. By then, any possible damage to workers has already been done, and if the legal level has been surpassed, the owner is subject to penalties.

To dispose of any hazardous material (here, even if the amount of asbestos is less than 160 ft^2 [*3*]), the abatement job still typically has to be registered with the proper state enforcement agency ten days before starting work. The asbestos must be properly handled and disposed of in a registered landfill; the creating of much paperwork is required. All of these rules are part of NESHAP [*4*].

If, during renovation or demolition, additional friable asbestos is discovered or created, this material also becomes subject to regulations, including the notification and disposal requirements.

In other words, if cement-asbestos board is broken during renovation or removal, it must be treated as friable. The broken edges must be coated with mastic, and the material sealed in plastic for disposal in an approved site [*5*].

Abatement procedure

Before starting the job, workers must have a full chest X ray and medical examination, including medical history and breathing-capacity test. Any smoking history is noted. The employer must maintain these records for 30 years and update them periodically.

To protect workers doing the job, full-body covering is required, such as disposable synthetic-nonwoven-fabric suits, two-cartridge respirators that employ high-efficiency particulate air (HEPA) filters, gloves and boots (see left photo).

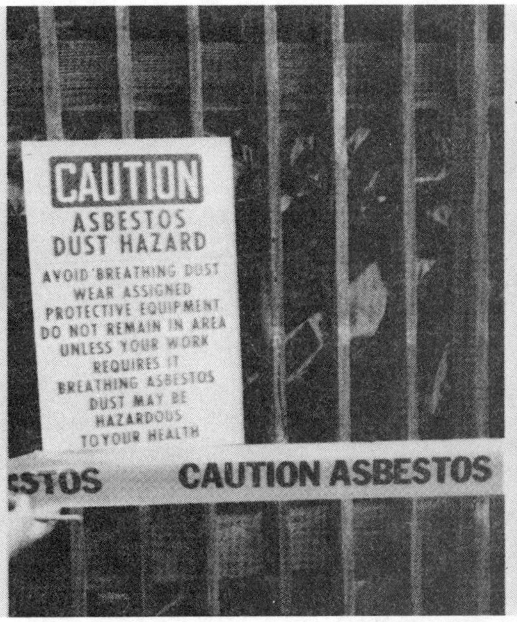

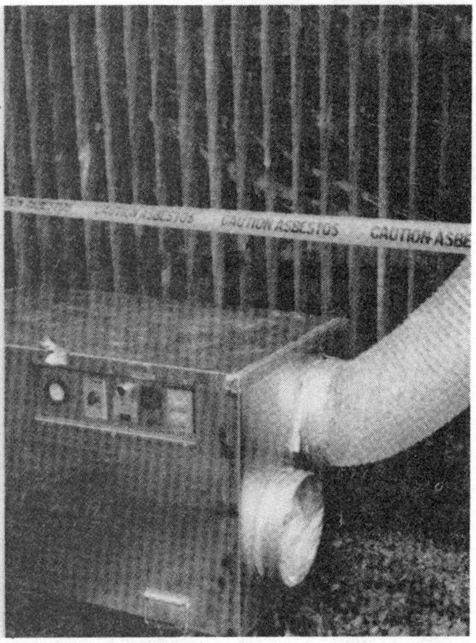

A decontamination unit must be at the site. The unit consists of a dirty room for work-clothing and equipment storage and entry to the work area, a shower room, and a clean room for the decontaminated worker.

Since other plant personnel will be curious about the work and the use of the "moon suits," the work area must be posted and roped off with yellow and black warning tape and printed hazard-warning signs (see center photo).

Protecting the environment

The environment is protected from asbestos fibers leaving the workplace by:

1. Using the decontamination room;
2. Employing a negative-air machine to filter workplace air through HEPA filters at a flowrate of about 1,950 cfm, depending on the size of the work area (see right photo); and

6-25

3. Sealing off the workplace with polyethylene sheet, 6 mils or more thick.

Disposal of the waste

The prime legal responsibility of disposal is with the facility that used the asbestos. Therefore, the operating company should make sure that its contractor is well versed in the procedures and has its workers certified by EPA. The contractor may need to hire a consultant, too. Knowing and following all of the rules, and keeping the proper paperwork is a critical part of the process (see top left photo).

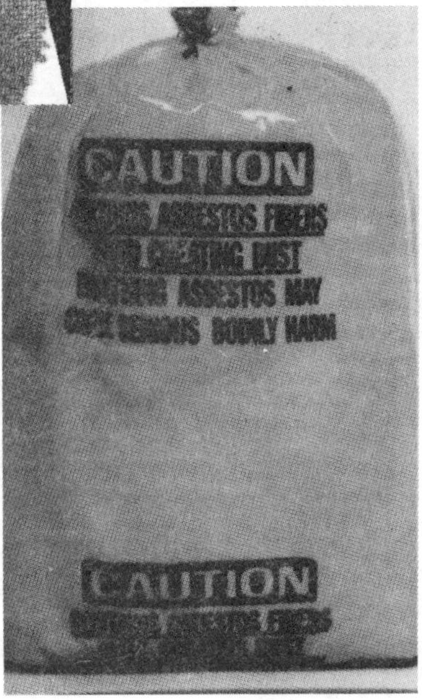

Some of the many forms required by the various government agencies are: employees' awareness medical examination form; asbestos disposal and documentation; visitors' documentation; daily checkoff list; worker respirator-training acknowledgment; owner prestart checklist; responsibility checklist; and supervisor, foreperson, safety and workers checklists.

To keep a minimum of loose fibers from escaping into the workplace, the asbestos is soaked in a water-surfactant mixture. The surfactant is usually a blend of polyoxyethylene ether and alkylated phenols.

The asbestos waste is put in special polyethylene bags, 6 mils or heavier, and each bag must have a warning legend on it (see bottom left photo). Before passing out of the dirty room, the bags are sealed and washed to decontaminate them.

Then they are put in a vehicle lined in polyethylene and taken to a Type 1 landfill (see bottom center photo). To contain any loose fibers, the vehicle is sealed before moving. Asbestos warning signs are posted on it.

Asbestos waste must be buried forever in a Type 1 landfill (see bottom right photo).

A chain-of-custody form must be passed from the waste-generating firm to the transporter and ultimately to the person who receives the toxic waste at the disposal site.

Penalties

Disposal in other than a Type 1 dump will incur federal fines of $25,000 per day until the situation is remedied. State fines are generally the same; local fines are mostly higher.

Penalties are levied against the owner of the facility where the asbestos came from and the contractor who did the dumping.

Hiring a consultant will help prevent such problems. A consultant can take samples inside and outside of the workplace to analyze the air for asbestos, both before and after the job has been done.

The law does not hold dump operators responsible for legal disposal. Therefore, at a Type 1 site, the owner, contractor or owner's legal representative must be present to verify that the asbestos is properly disposed of and buried.

CHAPTER 17

BID EVALUATION

When purchasing an automobile, it is quite simple to confirm the manufacturer's performance claims. All that is needed to check the acceleration (from 0 to 60 mph in 32 seconds) is a stop watch. Likewise, ascertaining the gasoline consumption is merely a matter of arithmetic.

The problem, however, in verifying the performance of a cooling tower is complicated since design criteria of operation (**Fig. 1**) is based upon gallons per minute of water circulation, water temperature on the tower, water temperature delivered by the tower, and wet bulb temperature (which consists of ambient temperature and relative humidity). These five variables are constantly changing, and the performance of the tower is heavily dependent upon the ever fluctuating wet bulb temperature.

Therefore, bids for new equipment, or rebuilding and upgrading of existing towers, must be carefully evaluated in order to obtain the best choice of equipment and to ensure reliability of production and overall lowest cost over a given number of years of service.

Selection Responsibility

To optimize the results, the selection responsibility should be shared by both the engineering department, which is familiar with the operating parameters of the cooling tower, and the purchasing department, which understands corporate requirements and has had considerable experience in contractor relations. Another useful addition to the bid evaluation team is a consultant or expert in the field whose specialized expertise is being called on more and more by industry today.

Basic Considerations

Three major components to be considered for solving the financial equation for the lowest total cost should include: capital expenditure; operating expense; maintenance expense.

The comparison between capital expenditure and maintenance charges can normally be figured, in a standard industrial operating situation, as a cost of \$1.00 of capital to 50¢ of maintenance, due to the tax deduction of maintenance as an operating expense. While the O.E.M. purchase price for new equipment can be capitalized, in many instances, the high cost of after tax dollars indicates a closer look at the rebuilding of existing equipment.

Prior to the energy crunch, which will be with us for a long time, the lowest overall price usually was considered regardless of the high cost of running the pumps, fan motors, and energy loads imposed upon the process for refrigeration equipment due to insufficient cold water. Since our current energy situation would suggest that the price of electricity and cost of energy will continue to increase markedly, consideration should encourage one to lean towards larger sizes which will increase original capital expenditure but would allow for lower air or smaller kilowatt consumption and colder water for better condensing temperature.

Another very important item to consider when deciding between rebuilding and purchasing new equipment is the high interest rate for borrowed money versus using budgeted before tax maintenance expenditures.

COOLING TOWER DESIGN CONDITIONS

1. GPM OF WATER VOLUME
2. ENTERING TEMPERATURE
3. LEAVING TEMPERATURE

WET BULB TEMPERATURE

4. RELATIVE HUMIDITY
5. AMBIENT TEMPERATURE

FIG. 1 — Design conditions describe the duty and work expected from a cooling tower. They consist of five variable input factors: water volume, temperature in, temperature out, and wet bulb temperature, consisting of relative humidity and ambient temperature.

Extremely low capital prime cost for a "cheap" cooling tower can turn out to be very expensive in the long run. On the other hand, maximum capital outlay vs. absolute minimized maintenance and operation costs may also lead to less than the optimum economic solution.

Thermal Performance Capability

With the foregoing financial items to consider, the most single important factor in cooling tower bid evaluation, without a doubt, is dependability of uninterrupted production of the unit involved. This is economically more critical than any one or combination of considerations in a capital expenditure of a cooling tower. Therefore, thermal performance capability of the cooling tower is the number one overriding requirement in order to achieve maximum uninterrupted unit production and consequent optimization of profits.

Performance consideration is all the more important by the verified condition of the industrial cooling tower market today. More and more purchasers of cooling towers are calling on the Cooling Tower Institute to conduct a strictly unbiased test of the tower capability they have purchased in accordance to the CTI ATP-105 Acceptance Test. This test is contracted to unbiased firms as the Midwest Research Institute. The best protection to obtain a cooling tower which will meet design performance criteria is to take care of properly writing specifications. Specify the CTI ATP-105 Acceptance Test with qualified observers, making certain that all prospective bidders understand that this test will actually be carried through with financial penalties for failure. Each bidder, whether it be for new equipment or rebuilding and thermal upgrading, should be informed that retention will not be released until the tower satisfactorily passes the performance test regardless of the time that has elapsed.

At this stage of bid evaluation, a number of manufacturers will claim they will meet the performance capability as specified. How then does an engineer approach the problem of which of these towers sized and built in such a manner that they will actually meet this performance? More often than not, the answer could be "why should we care? This is a reputable manufacturer and if for some reason the cooling tower does not meet the performance criteria set forth in the specifications, this manufacturer will make good."

Cost of Deficiency

In actual practice it has been found that even when a cooling tower shortage is corrected by the manufacturer, it is usually never done so without additional cost to the customer. There are most always involved additional basin, piping, and/or wiring costs. There are invariably escalated operating costs if total fan horsepower is increased. More important is the fact that the period between loading of the cooling tower and time the tower is actually brought into performance limits is usually lengthy, on the order of one or two years. Therefore, if the tower was properly sized to start with, which is in direct relationship to unit production, then unit production would not have been reduced one to two years by these performance shortages of higher operating temperatures which also entail higher levels of energy input.

A consideration of the financial loss involved in this reduction and crippling of production will in all cases far outweigh any benefits of manufacturer's reconstruction of the tower or the increased price that it would have cost to buy a higher performing tower in the first place.

Inquiry Form

As an assist comparing various manufacturers' or contractors' proposals, utilize the three page inquiry and bid form, Figure 4. By consolidating the submittals on one sheet, comparisons can be made as to energy pumping head, fan motor horsepower, size, and price. This comparison would make the best choice obvious.

Evaluation Items

Some of the specifics to look closely at in the bid proposals to determine performance adequacy are as follows:

- **SIZE** of tower, length, width, and height of box size.
- **HORSEPOWER** to move air through the tower. Watch in particular the size-horsepower relationship. The same performance can be obtained by building a smaller box with more horsepower, namely higher air velocities; or building a larger box and utilizing less horsepower, or lower air velocities. These are reasonably comparable performing cooling towers. However, if a tower presented to you is both smaller in size and utilizes less horsepower than other offerings, further investigation should be made of this proposal.
- **FILL** should be considered as to type on the basis of actual cubic foot volume rather than, for example, in a crossflow tower, the distance between the air intake louvers and drift eliminators. Distance between the louvers and eliminators to the fill section is significant; this determines the air travel through the fill packing. The fill spacing, both vertical and horizontal distances between the fill slats, and the size of the fill splash bars themselves should be considered. Often times this comparison will reveal one tower with considerably more or less fill than in other towers.
- Request **FAN** curves for the fan size involved and have the manufacturer submit supplemental data that the fans are not being utilized on his tower selection at near their stall point. If this condition exists, then any increase in the performance of the tower by increasing pitch of the blades or adding more horsepower will be useless since actually no additional air can be moved.
- **REFERENCE CHECKS** with owners of similar towers involved for reputation and ability to meet design performance criteria of the particular manufacturer or rebuilder involved are useful.
- **PUMPING HEAD** is an important evaluation item which can be calculated into horsepower, kilowatts, and consequent direct costs over the projected life of the tower. If is often neglected and it should be noted that some crossflow towers **(Fig. 2)** would require considerably more pumping head due to the design as compared to counterflow towers **(Fig. 3).** The table at the right indicates the dollar cost of additional horsepower pumping head. For example, the difference between a large crossflow and counterflow tower with 30′ of pumping head for 50,000 GPM of water will require an extra 450 HP. The table indicates that this will cost $185,888 the particular tower is operated, like most power generating stations, 8760 hours per year. The operating cycle of your tower can be calculated and the chart used to determine the extra horsepower penalty cost of additional pumping head. This dollar figure can be subtracted from the lower pumping head tower for lower fan horsepower utilization, thereby indicating that a tower that might cost a higher price initially will, in the long run on a maintenance and power and utilization basis, be a much better buy for your company.

FIG. 2 — Crossflow tower cooling 50,000 GPM of water will have a 60′ typical pumping head from the basin curb to the top hot water distribution area, requiring, in some installations, 950 HP to operate the pumps.

FIG. 3 — Counterflow tower cooling 50,000 GPM of circulating water will require an average of 30′ pumping head from the basin curb to the water distribution piping. In bid evaluation, the difference between counterflow and crossflow in charges for the extra 450 HP in the first year could be as much as $156,995 due to the extra pumping height.

• **EVAPORATION** loss is firmly tied to the specified design heat load conditions and amounts to essentially .1 of 1% of the circulating waterflow for every one degree of range to be cooled. There should be no appreciable difference in proposals on an evaporation loss listed on the bid form.

• **BASIN COST** is another tangible item that can be accurately evaluated by utilizing the dollars per year cost of concrete construction in the plant under consideration. Therefore, accurate basin size should be taken into account and evaluated on a capital cost basis. It is important to note whether or not a proposal includes basin costs.

• **PIPING** to the tower is important and the question of who will put in what pipes to where should be evaluated. Diameters must be looked into and how much pipe will be furnished by the customer and the cooling tower manufacturer. Balancing valves are an item as to which party will supply and install them.

• **MECHANICAL EQUIPMENT,** fan efficiency and speed reducers should be looked at. CTI service factor of 2.0 is standard. Shaft and coupling guards, together with vibration cut-out switch on each fan motor circuit are part of OSHA requirements (together with 2 access locations to top of tower, handrails, kneerails, toe boards).

• **FIREPROOFING** requirements for material would affect insurance costs and will vary as to sprinkler systems in the plenum chamber. Fireproof fills, self-extinguishing plastics, or wood are to be looked at.

• **MISCELLANEOUS** items should be scrutinized. Be certain that the price as quoted covers all costs of the tower. Those items that are not covered must be clearly understood. Such elements as receiving the material, unloading and storing of materials, utilities during construction of the tower, utilization of crane for construction, should all be called out in the proposal so that they may be included in the evaluation.

• **FINANCIAL** items such as the length of time the price is firm and escalation clauses in pricing should be considered against comparable escalation rates between the various bids. Terms of payments and discounts offered for prompt payment can be compared.

• **ENGINEERING BACKUP.** Another increasingly important item to scrutinize in today's pollution-conscious world is the ability of the cooling tower vendor to furnish engineering help and solutions to make a cooling tower design compatible with all applicable government rules and regulations in the area under consideration such as quantity and quality of blow down, drift loss, air pollution etc. Availability and promptness of service on the equipment once it is put on the line from the manufacturer is also of prime concern.

In Conclusion

1. **DO NOT** consider the purchase of a cooling tower on basis of low dollar only.
2. **DO** evaluate a complete cooling tower proposal from engineering and purchasing personnel standpoints.
3. **ENSURE** that the vendor fully understands that a certified acceptance test in accordance with CTI ATP-105 will be performed with a retention to be held until the tower passes the specified level of capability.
4. **DO** examine the details of materials and construction on a comparable basis and in accordance with CTI specification standards.
5. **DO** fulfill your obligation to your company by purchasing the optimum selection for the most overall economic tower that will maintain uninterrupted dependable unit production. □

TABLE I. Electric Motor Operating Cost

Motor Size HP	Full Load *KVa* @ 1800 RPM	Operating Cost Per Year (8760 hours) at Stated Cost per kWh ($)							
		.03	.04	.05	.06	.07	.08	.09	.10
10	10.14	$ 2,665	$ 3,553	$ 4,441	$ 5,330	$ 6,218	$ 7,106	$ 7,994	$ 8,883
15	15.67	4,118	5,491	6,863	8,236	9,609	10,982	12,354	13,727
20	20.06	5,271	7,029	8,786	10,544	12,301	14,058	15,815	17,573
25	24.93	6,551	8,735	10,919	13,103	15,287	17,471	19,655	21,839
30	28.58	7,511	10,014	12,518	15,022	17,525	20,029	22,532	25,036
40	38.32	10,070	13,427	16,784	20,141	23,498	26,855	30,211	33,568
50	47.10	12,378	16,504	20,630	24,756	28,882	33,008	37,134	41,260
60	57.21	15,035	20,046	25,058	30,070	35,081	40,093	45,104	50,116
75	71.10	18,685	24,913	31,142	37,370	43,599	49,827	56,055	62,284
100	91.11	23,944	31,925	39,906	47,887	55,869	63,850	71,831	79,812
125	118.47	31,134	41,512	51,890	62,268	72,646	83,024	93,402	103,780
150	140.57	36,942	49,256	61,570	73,884	86,198	98,511	110,825	123,139
200	188.65	49,577	66,103	82,629	99,154	115,680	132,206	148,732	165,257

COOLING TOWER BID FORM

Sheet 1 of 3

Customer	Manufacturer
Name	
Firm	
Address	
Request No. Date	Job No.

____ Counterflow
____ Crossflow
____ Type ____________
____ Model Number ____________

SUMMARY

A. DESIGN SPECIFICATIONS	MFG 1	MFG 2	MFG 3	MFG 4
Water Flow ~~vs~~ is GPM				
Inlet Temperature HWT°F				
Outlet Temperature CWT°F				
Wet Bulb Temperature WBT°F				
Elevation, ft.				
Evaporation Loss %				
Drift Loss % Flow				
Total Fan BHP				
Tower Pump Head				

look at [illegible]

B. PHYSICAL DATA	MFG 1	MFG 2	MFG 3	MFG 4
Number of Cells				
Fans Per Cell				
Cell Dimensions L x W, ft				
Total Dimensions L x W, ft				
Fan Cylinder Height, ft				
Basin Depth				
Tower Height (less stack) ft				
Riser Pipe Diameter, in.				
Basin Curb to Header, ft.				
Shipping Weight, lbs.				
Operating Weight, lbs.				

COOLING TOWER BID FORM

	MFG 1	MFG 2	MFG 3	MFG 4
C. MATERIALS				
Siding				
Structure (RW, DF)				
Wood Chemical Treatment				
Structure Connectors				
Bolts, Nuts, Washers				
Nails				
Anchor Connectors				
Wet Decking				
Drift Eliminators				
Air Intake Louvers				
Distribution Piping				
Spray Nozzles				
Fan Deck				
Cold Water Basin				
Mechanical Equipment Supports				
Handrails, Kneerails, Toerails				
Stairway				
Ladder				

	MFG 1	MFG 2	MFG 3	MFG 4
D. MECHANICAL EQUIPMENT				
1. Motor				
Quantity				
Nameplate H.P.				
Total BHP Per Cell				
Manufacturer				
Type				
Kind				
Electric Characteristics				
Operating RPM				
Service Factor				

	MFG 1	MFG 2	MFG 3	MFG 4
2. Drive Shaft				
Quantity				
Type				
Manufacturer				
Model Number				
Shaft Material				
Coupling Material				
Rated Horsepower				

COOLING TOWER BID FORM

	MFG 1	MFG 2	MFG 3	MFG 4
3. Gear Reducer				
Quantity				
Type				
Manufacturer				
Model Number				
Rated Horsepower				
Reduction Ratio				
Number of Reduction				
4. Air Mover				
Quantity				
Type				
Manufacturer				
Model Number				
Diameter				
Number of Blades				
Tip Speed, FPM				
Fan Speed, RPM				
Air Volume, ACFM				
Total Static Pressure H_2O				
Static Efficiency				
5. Thermal Data				
Liquid to Gas L/G				
Characteristic KaY(V)/L				
E. QUOTATION				
$				

Tower shall pass a certified CTI-ATC-105 test

10% retention shall be withheld until tower meets contract specifications

contractor will do whatever is necessary to bring tower to design specs & pay for new certified test

CHAPTER 18

FIELD INSPECTION

PROFESSIONAL FIELD ENGINEERING INSPECTION

Previous chapters have indicated that cooling tower upgrading is not a hit or miss proposition, but a definite Engineering discipline based upon thermal mass transfer, aerodynamics, and hydraulics. Operators are becoming more aware that the limiting factor in production and profits could be the performance of the tower.

Day to day operating and maintenance inspections should be performed by the personnel involved on a periodic basis. However, a thorough inspection by a professional cooling tower field engineer is strongly recommended for the long range economic investment.

Professional inspection will indicate hidden flaws and weaknesses which might be overlooked by the operating personnel. The inspector will also outline in a thorough manner all of the maintenance and repair items that are necessary together with budgetary recommendations for their implementation, together with the physical condition of the structure and the mechanical equipment.

All of this will be performed with the inspection sheet similar to the "punch List" on page 137.

The cooling tower fact sheet on the following page indicates the type of data gathered for an engineering study of the thermal condition of the tower. The inspector will examine the actual operating logs to determine the present energy consumption of the system and tie this in with a certified test, if available, to ascertain the level of efficiency of the tower.

With the new required Design Conditions, based upon available air, he can then calculate if a significant thermal improvement is available, and to what extent. Part of his report should indicate the period for the cost-effective payback for rebuilding the tower thru energy savings due to having colder water available for the chemical process or machinery cooling.

In these days of expensive real estate and construction costs, it is quite possible to more than double the capacity of many cooling towers by a modern engineering rebuild, which also should be commented on in the inspection report.

The information requested on the fact sheet, of course, can also be furnished by in-plant personnel to determine, on a preliminary basis if upgrading is feasible. However, the relatively small cost of a professional inspection is a good investment in future returns of profits amd energy conservation.

Requested by: ______________________

Name, Title ______________________

Company ______________________

Address ______________________

____

Phone: ______________________

____
RBA Job No.

Location ______________

Cooling Tower Fact Sheet

Supply as much data as available

Date

Cooling Tower Manufacturer ______________ Model Number ________ Number of Cells ________

Counter Flow ☐

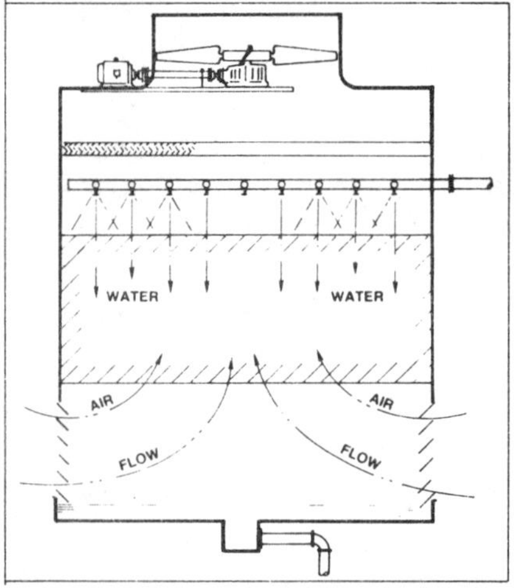

Cross Flow ☐

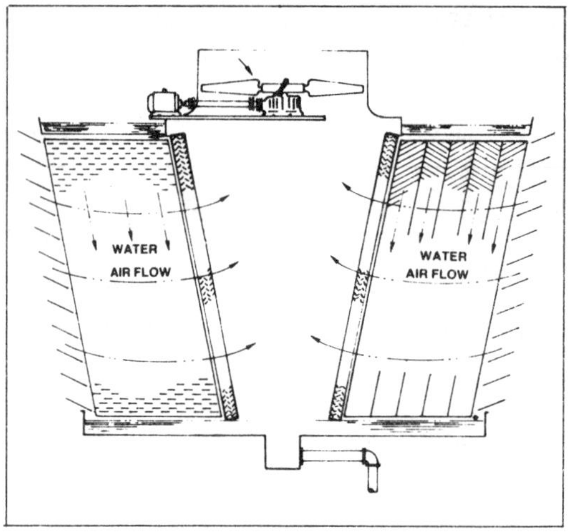

Fan Diameter (Feet) ________

No. of Fan Blades/Fan ________

Fan RPM ________

Fan CFM ea. ________

Motor HP ________

Motor RPM ________

Gear Reducer Ratio ________

Riser Pipe ∅ ________

Original Design Conditions ☐

Present Operating Conditions ☐

Gallons Per Minute of Water:

Each Per Cell ______________

Total System ______________

Temperatures: Water:

On Tower ______________°F

Off Tower ______________°F

Ambient Wet Bulb ______________°F

Tonnage of Refrigeration ______________

Design Conditions Desired

Gallons Per Minute of Water:

Each Per Cell ______________

Total System ______________

Temperatures: Water:

On Tower ______________°F

Off Tower ______________°F

Ambient Wet Bulb ______________°F

Tonnage of Refrigeration ______________

DIMENSIONAL INFORMATION

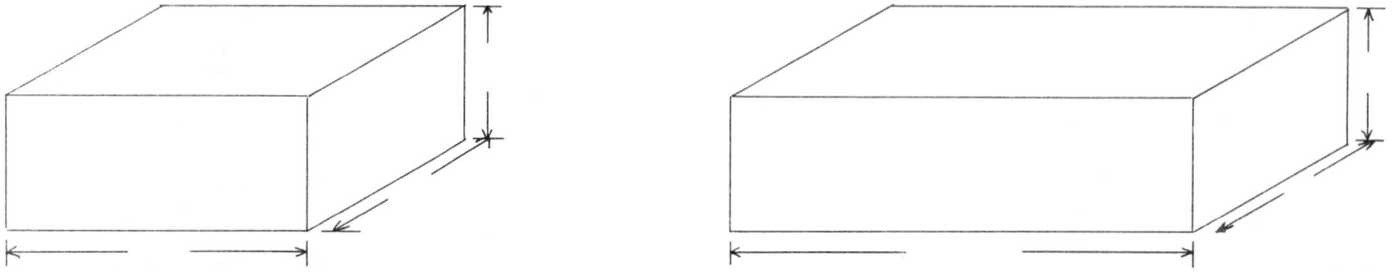

Length ________ Width ________ Height ________

Indicate fan position(s) on applicable sketch

CHAPTER 19

REDWOOD IN COOLING TOWERS

REDWOOD IN COOLING TOWERS

Cooling towers fabricated from Redwood lumber are very unique structures compared to other wood assemblies, such as houses. In cooling towers there are more columns by upwards to 50% of nominal 4 x 4 lumber than there are frames in comparable construction. Considerable amounts of these columns under normal operating conditions can conceivably be loaded near the limits of their maximum allowable design load.

Standard design parameters for practically all structures are for live loads higher than actually encountered. This provides for an additional factor of safety which is not usually the case for cooling towers. The primary design live load is water and since the unit weight of water is quite well defined, the heavily loaded columns in the cooling tower are actually subjected to their design load.

Another factor causing concern and requiring constant inspection of timber columns is that in cooling tower usage, the structural members are exposed to a hostile environment. The temperature ranges can vary from 105° in airconditioning requirements upwards to 150° for process water. High humidity and saturated environment are normal and the timbers are subjected to cycles of wetting and drying. These factors will affect the properties of the lumber and should be under constant surveillance for deterioration.

Careful and competent inspection must be maintained with vigilance to guard against premature failure of weight and stress bearing members causing shut down and requiring extensive repairs.

California Redwood has played a very important role in the Cooling Tower Industry since it has been used successfully in the construction of these structures from the turn of the century. According to the California Redwood Association (C.R.A.), if you took all of the Redwood that has gone into cooling towers alone, you could create a plank of wood one inch thick and one foot wide that would stretch around the world 7½ times amounting to approximately one billion board feet, as a conservative estimate.

The most important fact making Redwood a desirable material for cooling tower construction is that it costs less in the long run than any other material. Besides being readily available, its history of successful service is important and compared to other materials, it is easy to work. The Redwood members can be completely prefabricated in the plant to control labor costs and avoid the variables of on-the-job site construction.

The inherrent properties of Redwood are important considerations in using this material since they have performed more satisfactorily under the extreme conditions found in cooling tower operations. Durability is foremost for its recommendation. The native ability to resist the deteriorating effects of hostile elements is a measure of the durability of the wood. Decay, biological attack and the errosive effects of weather are the most common enemies of the wood. A wood that can withstand these antagonists, by maintaining its original shape and strength, is considered high in durability properties. Among commercial species, Redwood is unexcelled in natural durability. Inherrent chemicals which give the clear heartwood of California Redwood its rich, ruddy color, also repel insects and decay producing fungi. These natural repellants are in every cell and fiber so that clear heart Redwood is specified for construction where conditions require decay and insect attack resistance.

The third important property of Redwood is its dimensional stability or lack of shrinkage. Of any commercially used wood in America, Redwood has the lowest volumetric shrinkage. Its dimensional stability is maintained to a high degree since it has relatively little tendency to warp or check. Since dimensional stability is maintained to a high degree, and so therefore, is the structural integrity of its application.

Chemical resistance of Redwood is high. The chlorine content and pH of cooling tower waters is an important consideration that is resisted by the Redwood Composition. The natural resistance to disintegration by chemical solution or their fumes, even at elevated temperatures, enabled Redwood to become an important material for many Industrial uses.

California Redwood is available in many lumber grades which will be enumerated in descending order of desirability.

CLEAR ALL HEART is strictly heartwood and the highest quality available which is often used in tower construction. It is free of all lumber characteristics such as knots and it is unexcelled in dimensional stability and resistance to weather erosion. This grade meets the highest standards for industrial uses as cooling towers, potable tanks and pipes, scrubber towers, degasifiers. *Figure 1.*

Figure 1 – Clear All Heart Redwood is the highest quality available.

SELECT HEART is the next heartwood-only grade. This classification is free of characteristics such as shakes and splits. Sound tight knots are allowed and may be up to 1/3 the width of the piece of lumber. *Figure 2.*

Figure 2 – Select Heart Redwood grade is free of shakes, pits, but permits sound tight knots.

CONSTRUCTION HEART is the lowest recommended grade of heartwood available that is specified in cooling towers. This classification permits larger knots and other characteristics yet it is still quite suitable for structural framing with only a slight decrease in strength and offering a significant increase in economy. *Figure 3.*

Figure 3 – Construction Heart Redwood is the lowest recom-

Pressure and chemical treatment of the wood is usually specified for additional protection against attack and deterioration, but this treatment does affect the grade classification of the lumber.

Chemical treatment of Redwood to extend its useful life by upwards to 50% against chemical, fungal, and decaying attack normally consists of impregnation with Arsenic Acid and Copper Chromate in solution. (A.C.C.)

Pressure treatment of this type for Douglas Fir is in considerable use today since the chemical treatment will act as a long range preservative of the Fir. Most cooling tower manufacturers tend to prefer the Redwood since Redwood untreated is highly superior to untreated Douglas Fir. It is reasonable to assume therefore, that the same chemical treatment of both materials will be considerably in the favor of the use of Redwood.

Douglas Fir became popular in use in the 1940's following the difficult availability of Redwood directly after World War II, and during prolonged labor difficulties in the Redwood forests. However, in today's market, there is sufficient Redwood availability in the market place so that substitutes are not necessary. The price advantage of pressure treated Douglas Fir is not sufficient, on a practical basis, to use this material as a money saving substitute.

The following commentary was obtained through the courtesy of the California Redwood Association in response to critics who claim that Redwood is an endangered species, that Redwood availability is rapidly decreasing and that our Redwood forest resources are being depleted.

> "First, let us distinguish between two types of Redwood species: the Sierra Redwood, and the Coastal Redwood. The Sierra Redwood is the brute of the family. The giant Sequoia is the largest of all living things on earth. These trees are far heftier than their coastal cousins. But they are also shorter. Today, 99% of them are preserved in Federal and State Parks, stretching some 260 miles from a point near Lake Tahoe, south to Kings Canyon National Park.

The California coastal Redwoods are the source of the timbers, fill, and siding used for cooling towers. These are the tallest trees on earth. They are found hugging the California coastal region, from a few miles north of the Oregon border to a point of 500 miles south, almost to the famous Hearst Castle at San Simeon, California. These remarkable trees provide us with a perpetual supply of raw material, amounting to a production volume of one billion board feet annually. And that is where it will remain. Because modern, responsible forest management practices exercised by the Redwood Industry in cooperation with a number of private and governmental agencies are making sure we will never run out of Redwood timber. That means that Redwood products will always be available.

Admittedly, logging practices – and our forests – have come a long way from the irresponsible cutting that alarmed the Nation during the early part of this century. Lands that were completely logged off more than 100 years ago now have stands of Redwood that are probably the most dense, most heavily producing Redwood forests anywhere on earth. In fact, the Society of American Foresters believes that there are more Redwoods growing today than when man first started cutting them. We have seen the growth of young Redwood trees in commercial forests double since World War II. And it is expected to double again by the year 2000.

There is a serious misconception that Redwoods are particularly slow to grow, and take a long, long time to mature. The fact is that Redwood is the fastest growing commercial Conifer in the country. And it is virtually indestructible. The coastal Redwood has been there for 40 million years. Primarily because it reproduces itself by seeding, as well as by sprouting from the stump and root system of parent trees. That is something no other softwood species can do on a commercial basis. When a single tree is cut, as many as a dozen or more new trees spring up from the roots. In 40 to 60 years, between 3 and 5 of these sprouts will stand 100 to 150 feet tall and measure two to five feet in diameter.

The Redwood Industry is seeing to it that our highly productive Redwood forests are not merely surviving, but flourishing. Today, more than 134 tree farms are committed to sustained yield by modern forest management, which operate on 949,700 acres. These lands are rich in timber, fish, wildlife, and clean water. They are protected from fire, disease, insects and other environmental damage. Another 181,000 acres of our tallest, largest and oldest coastal Redwoods are preserved in Federal and State Parks and public reserves. That is 283 square miles of untampered forests available for public recreation.

Timber harvests are being carefully planned and implemented in strict accord with California's Forest Practice Act 1973. This legislation is the most stringent and restrictive control ever enacted to regulate timber harvesting on private property. The act demands that each logging plan be accompanied by an appropriate reforestation plan, according to sustained yield management practices.

Experts are predicting that at the end of this century our need for Redwood products will double. How can the Redwood Industry remain optimistic, faced with pressure from Environmentalists to cut less wood on one hand, while on the other hand we find growing demands for more and more wood utilization?

The answer lies in the exciting results of forestry research, and the industry's willingness to invest millions of dollars to implement that research. One recent example of this kind of investment is this brand new tree nursery near Korbel, California. The largest of its kind in the Redwood region, this nursery will eventually product 4.5 Million seedlings annually. Special seedlings with a few improvements on Mother Nature's genetic plan. These super seedlings will produce straighter, larger, taller trees that will grow faster than ever, with a

95% survival rate. Soon we will be able to grow a 60 year old size tree in just 40 years. Improved forest management practices now support these fascinating genetic improvements. Carefully planned thinning procedures free young trees from crowding, thereby producing up to 10 times faster growth. This practice also increases ground foliage and wildlife. Seeding by helicopter, planting seedlings by hand, and soil fertilization are now standard procedures for modern reforestation."

INSPECTION OF WOOD COOLING TOWERS

Cooling tower maintenance and repairs are part of the cost of living today, and a scan of any day's newspaper will turn up a reminder that the cost of living continues to rise. The word we least like to hear is *expensive*.

We can however, minimize the impact of this word on cooling tower repairs by making comprehensive inspection a regular item of the maintenance program.

Mechanical components of a tower are the pampered children – receiving close attention, scheduled feedings and prompt changes, as required. But, what about the fill, drift eliminators, distribution system and the structure? Too often, deterioration has advanced to the critical stage, in these areas, before attracting deserved notice.

Unforeseen repair or replacement may be required at a most inopportune time, resulting in a non-scheduled shut-down and far more expensive than the actual repair itself.

Wood cooling towers – even with pretreated lumber – should be given a thorough inspection on a regular basis: Annually for towers with over ten years service and at two to three year intervals for newer units.

In addition to mechanical equipment, fill condition and water distribution should be inspected at all levels of the tower. Tower lumber should be examined for evidence of fungal activity, with a record of the findings maintained to establish the *rate of deterioration* through the years.

Obviously, the most comprehensive inspection can usually be provided by Firms specializing in cooling tower manufacture and rehabilitation. But, operating, maintenance, or engineering personnel on the plant could perform this function with a reasonable degree of efficiency – time and manpower permitting.

Now, what to look for? There are several species of wood destroying fungi, but we can simplify to the three most commonly found in the cooling tower environment:

1. Soft rot, or delignification
2. Brown cubical, or iron rot
3. White pocket rot

Soft rot *(Figure 4)* will occur throughout the tower's flooded and vapor areas. It is characterized by a loosening of surface fibers and, in advanced stages, wood members noticeably are reduced in cross-section and strength. Thinner members, such as fill splash bars and eliminator slats, may be observed to sag considerably between supports, becoming very brash and darkly discolored internally. Framing timbers can be checked by scratching the surface with a screwdriver blade, to determine the depth of deterioration. Soft rot usually advances at a relatively slow rate and can often be substantially retarded by the application of wood preservative chemicals.

Figure 4 – Soft Rot loosens surface fibers, reduces cross-section and strength.

Brown cubical *(Figure 5)* will be found in close proximity to rusted iron or steel in direct contact with the wood, and the brownish stain on the surface is often the first clue. In advance stages this activity destroys the cellulose leaving a brown or black, crumbly or powdery lignin pocket with severe loss of member strength. Suspected areas may be probed with a screwdriver blade, and easy penetration coupled with the appearance of charred wood will tell the story of advanced rot. Note: Probe with the screwdriver blade *across* the grain – not with, or parallel to the grain.

Figure 5 – Brown Cubical Rot occurs in close proximity to rusted ferrous parts.

White pocket rot *(Figure 6)* can be the most devastating wood destroyer to be encountered. It is certainly the most deceptive, in that it is not apparent on visual examination until the wood member under attack has completely failed. This fungus, although like the other in that the initial assault is in the form of airborne or waterborne spores, thrives essentially in the interior of the wood – leaving the surface appearance virtually unchanged.

Figure 6 – White Pocket Rot is fungal attack on interior lignin cellulose bond.

Inspection for white pocket rot should be done by sounding the wood with light hammer blows, comparing the variations in sound between healthy and infected members. Good timbers, when struck, will have the characteristic 'ring' of new lumber, whereas the presence of fungal activity will be signalled by a rather dull 'thud' and, in advanced stages of pocket rot, the hammer will completely penetrate the stick. Essentially, pocket rot destroys the lignin, which is the bond between cellulose fibers, resulting in a stringy, porous mass beneath the surface.

Normally, white pocket rot will be more or less confined to the vapor or plenum area of the cooling tower, and inspection for this type of attack would be from drift eliminators to fan discharge point.

During the inspection, the condition of bolting and other hardware items should be noted, as well as the appearance of steel and cast parts of the tower. Be alert for clogging of distribution nozzles and drift eliminators which can reduce the tower efficiency. Look for unusual variance in clearance between a given fan blade and the interior circumference of the fan stack.

There are some of the fundamentals of tower inspection which may be employed by users to maintain an awareness of the physical profile of wood structure cooling towers. Certainly, as the occasion warrants, the services of manufacturers and experienced tower rebuilders should be enlisted to verify the need and extent of repairs, as well as to profit from these experiences and fingertip knowledge of the most modern materials and methods in use in a continually changing industry.

The cooling tower is a vital component of the system it serves, and its operating efficiency can be directly translated into dollars. Any machine will require periodic repair, but the capability of accurately predicting the scope of repair and controlling the time schedule is a prime economic advantage. Regular and detailed inspection of the equipment is the means towards this end.

Corrosion in wood frame cooling towers is minimal since there normally is not much steel in these types of structures.

However, there are certain areas to investigate, namely the mechanical equipment, mechanical equipment supports, timber and wood fasteners, together with anchor castings.

If these items are fabricated from carbon steel, regardless of the corrosion protection – be it either galvanic coating or paint – they should be investigated carefully for corrosion. The author suggests that the chapter on corrosion of steel frame towers be reviewed together with attention paid to the surface protection, priming, and coating of any carbon steel present in the tower.

While it is common practice to use heavy timbers to support the mechanical equipment, if steel beams are used, they should be carefully examined for corrosion since the effect of oxidation will reduce the strength of the supports and could lead to serious deterioration and/or replacement costs.

CHAPTER 20

REBUILDING FOR PROFIT

How cold water Conserves energy, Aids profit picture

■ In many instances, it is doubtful that managers consider (or are aware of) the role of cooling towers in the profit or loss picture. It is the operating engineers' function to utilize equipment as efficiently as possible to produce profits for management.

The first and major cost to consider is the electrical and/or steam dollar expenditure to power the systems. The cooling tower is a high energy utilization machine as well as a mass heat transfer device.

The second area of profit to analyze is the utilization of cold water. In chemical product manufacturing plants (Fig. 1), the colder the water, the more efficient the condensation process and the greater the volume of product yielded at lower cost. In refrigeration/air conditioning equipment (Fig. 2) the colder the water, the less energy required to power turbines and compressors. In power generation the hot water energy penalty deprives the facility of additional power sales.

After the cooling tower has been erected, proper operation and maintenance is required to yield full unit production. Towers can lose appreciable performance as soon as they are turned on when not properly maintained. To emphasize the actual dollar and cents loss contributed by a malfunctioning cooling tower, the following calculations are reprinted with the permission of James L. Willa,

Fig. 1 — Chemical plant uses cooling tower water to produce salable condensate product.

Loss of thermal performance of cooling towers is unnecessary, and often occurs in systems where cooling capacity is the limiting factor in high unit production.

Once the tower is installed, it must be operated and maintained properly in order to accomplish full unit production. Improper maintenance means loss of performance which can amount to as much as 5° added to cold water temperature. What does 5° in cold water temperature from a cooling tower mean to an operating unit? First let us consider a petrochemical plant unit with an overhead condenser operating at a pressure to condense all components over the temperature range 140°F - 80°F. For simplicity, assume straight line condensation.

Assume:
Heat of vaporization = 120 BTU/#
Specific heat of the vapor = 0.4 BTU/#/°F

Assume:
Total inlet flow = 460,000 #/hr Hydrocarbon

Then:
Q = (460,000) (.40 x 60°) + (120°) = 66.24 MMBTU/hr

Assume: U = 200
Area = 13,500 ft²

% Rec.	#/hr Cond.	Outlet HC Temp, °F	T HC	MM BTU/hr Sensible	MM BTU/hr Latent	MM BTU/hr Total
100	460,000	80	60	11.04	55.20	66.24
90	414,000	86	54	9.93	49.68	59.61
80	368,000	92	48	8.83	44.16	52.99
70	322,000	98	42	7.73	38.64	46.37
60	276,000	104	36	6.62	33.12	39.74

Assume:
unit is designed for 80% recovery
using 30° range on cooling water.

Required MTD = $\dfrac{52.99}{200 \times 13{,}500}$ = 19.64°F

Cooling Tower Design

Q	=	52.99 MM BTU/hr
GPM	=	3533
R	=	30°F
WB	=	70°F
CWT	=	80°F
HWT	=	110°F

Calculated Design MTD for operation of condenser = 19.64°F

Consider:
Operating tower in poor repair at 85°F CWT

By trial-and-error calculations:
Heat removal = 48.577 MMBTU/hr
Outlet Hydrocarbon Temp. = 96°F
Hydrocarbon condensed = 0.337 MM#.hr
Product Loss = 368,000 - 337,000 = 31,000 #/hr = 8.4%
Assume value = $0.01/#
loss/hr = 31,000 x .01 = $310/hr = $7,440/day

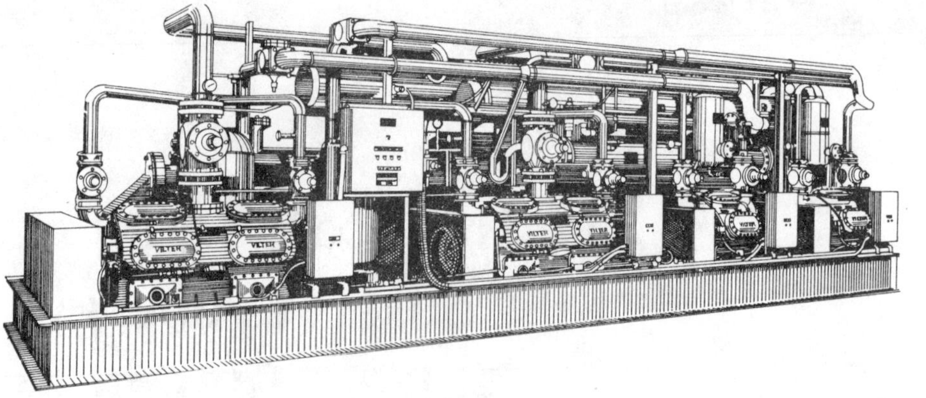

Fig. 2 — Refrigeration plant uses cooling tower water to operate at lower head pressures and energy input.

This example indicates an 8.4% differential in product recovery from the overhead condenser and a dollar cost which could easily pay for the cooling tower involved in less than two weeks.

Let's go through this same exercise on an air conditioning unit. Consider a 590-ton absorption type machine. Again, we assume a 5° rise in cold water temperature due to poor operation and maintenance conditions. In this case we will assume a normal Gulf Coast air conditioning cold water temperature design of 85°F and a 90°F cold water in actual operation on a 2120 GPM cooling tower. Assume operation at 44°F chilled water temperature, and 12 PSIG steam pressure. Typical manufacturer's data for this situation indicates a loss in capacity of 72 tons (12.2%) due to the high condenser water temperature. The incremental cost of equipment to increase tonnage back to design is approximately $500/ton for this type of operation. Thus the total cost differential is

$500 x 72 = $36,000

which is close to the original cost of the cooling tower.

Let us consider a 600 megawatt steam turbine power plant, operating at 440 megawatts with a heat rate of 7946 BTU/KW.

An increase in condensing pressure from 3.0″ to 3.5″ Hg absolute (115.06°F to 120.56°F) will increase the heat rate by 0.67%. That is, the *added* heat rate is .0067 x 7946 = 53 BTU/KW.

Assume fuel cost is $1.00/million BTU.

For 400 megawatt production, the added cost of the 5.5°F increase in condenser water temprature is:

$53 \times 440 \times 10^3 \times \$1.00 = \$23.32/\text{hour} = \$204{,}300/\text{year}$

Alternatively, in lieu of added fuel cost, consider lost production:

For the 5.5°F increase in condenser temperature,

Capacity reduction = 0.68%

$$\text{KWH loss} = \frac{440 \times 1000 \times .68}{100} = 2992 \text{ KWH}$$

Representing sale price of electricity (for home use) = 3.6¢/KWH.

$$\text{Loss/day} = \frac{2992 \times 3.6 \times 24}{100} = \$2{,}585 = \$944{,}000/\text{year}$$

The aforementioned example then shows a $204,300 per year increase in fuel costs to maintain production due to the 5°F performance shortage in the cooling tower, or a loss in retail electricity of $944,000 per year maintaining the same heat rate.

Note that the text uses a 5°F performance loss in each example. While this reduction of efficiency is quite high, it is not uncommon. However, the following, on a straight line proportion, indicates the dollar volume cost for relatively less performance which might be applicable to the reader's current operating situation.

OVERHEAD CAT CONDENSER

Product Less/day/°F Tower Defficiency

5° warmer water cost	$7,440.00 day
4° warmer water cost	$5,952.00 day
3° warmer water cost	$4,464.00 day
2° warmer water cost	$2,976.00 day
1° warmer water cost	$1,488.00 day
½° warmer water cost	$ 744.00 day

AIR CONDITIONING 590 TON ABSORBER SYSTEM

Additional Energy cost per year

5° warmer water requires	$36,000.00 year
4° warmer water requires	$28,800.00 year
3° warmer water requires	$21,600.00 year
2° warmer water requires	$14,400.00 year
1° warmer water requires	$ 7,200.00 year
½° warmer water requires	$ 3,600.00 year

600 MEGAWATT POWER PLANT

Capacity and Energy Penalty per year

	ENERGY PENALTY	RETAIL LOSS
5° warmer water cost	$204,300.00 year	$944,000.00 year
4° warmer water cost	$163,440.00 year	$755,200.00 year
3° warmer water cost	$122,580.00 year	$566,400.00 year
2° warmer water cost	$ 81,720.00 year	$377,600.00 year
1° warmer water cost	$ 40,860.00	$188,800.00 year
½° warmer water cost	$ 20,430.00	$ 94,400.00 year

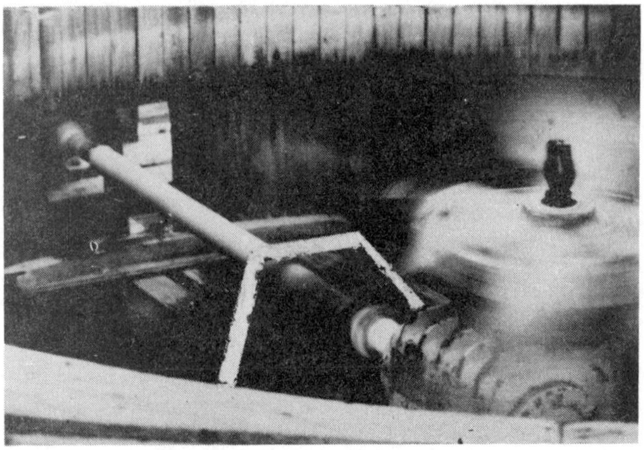

Fig. 4 — High cost of tower being inoperative due to fan wreck can be prevented by simple shaft guard installation.

Fig. 5 — Cellular fill is a high heat transfer medium which returns dollars to the owner by producing up to 40% colder water (depending upon conditions).

These statistics should indicate to the operating enginer his importance to overall profit of management.

Application of Cooling Tower Technology

Now using the information we have covered as to the various elements, let us go to examples to see how the application of modern cooling tower technology has saved money, and earned rapid pay back and profits for management.

Fig. 16: Rebuilding of the left counterflow tower for "cat" cracker at oil refinery permitted an additional 3,000 GPM to be used in process cooling.

Oil Refinery Counterflow

This Indiana cracking facility increased its output and required an additional 3,000 gallons of water per minute for the process equipment. The options were to install an additional cells on the counterflow tower (Fig. 16), or divide the new load between the two existing counterflow and crossflow towers.

Investigating the problem from a rebuilding engineering point of view, it was determined that if the large, inefficient, triangular splash bars were replaced with 42 inches of cellular fill, and the three 50 horse power motors rewound for 60 horsepower duty, they would generate sufficient air to cool 4,160 gallons per minute of water each per cell, which could then be cooled to the design requirements of water entering the tower at 102°F, cooling to 85°F at an ambient wet bulb temperature of 77.5°F.

One of the keys to making this upgrading work was the replacement of the gravity water trough distribution to a PVC piping system, used in conjunction with the large orifice ceramic nozzles. As an extra margin, new cellular drift eliminators were also installed.

Department Store Crossflow for Air Conditioning

Fig. 15 is a crossflow tower which was not producing cold enough water: the wood fill was deteriorating, steel corroding, and the Fire Insurance underwriters requested an automatic chemical deluge system be installed. The rebuilt tower at $68,000.00 completely fulfilled the contract in all specifications. Comparisons are listed to illustrate the difference in cost of rebuilding.

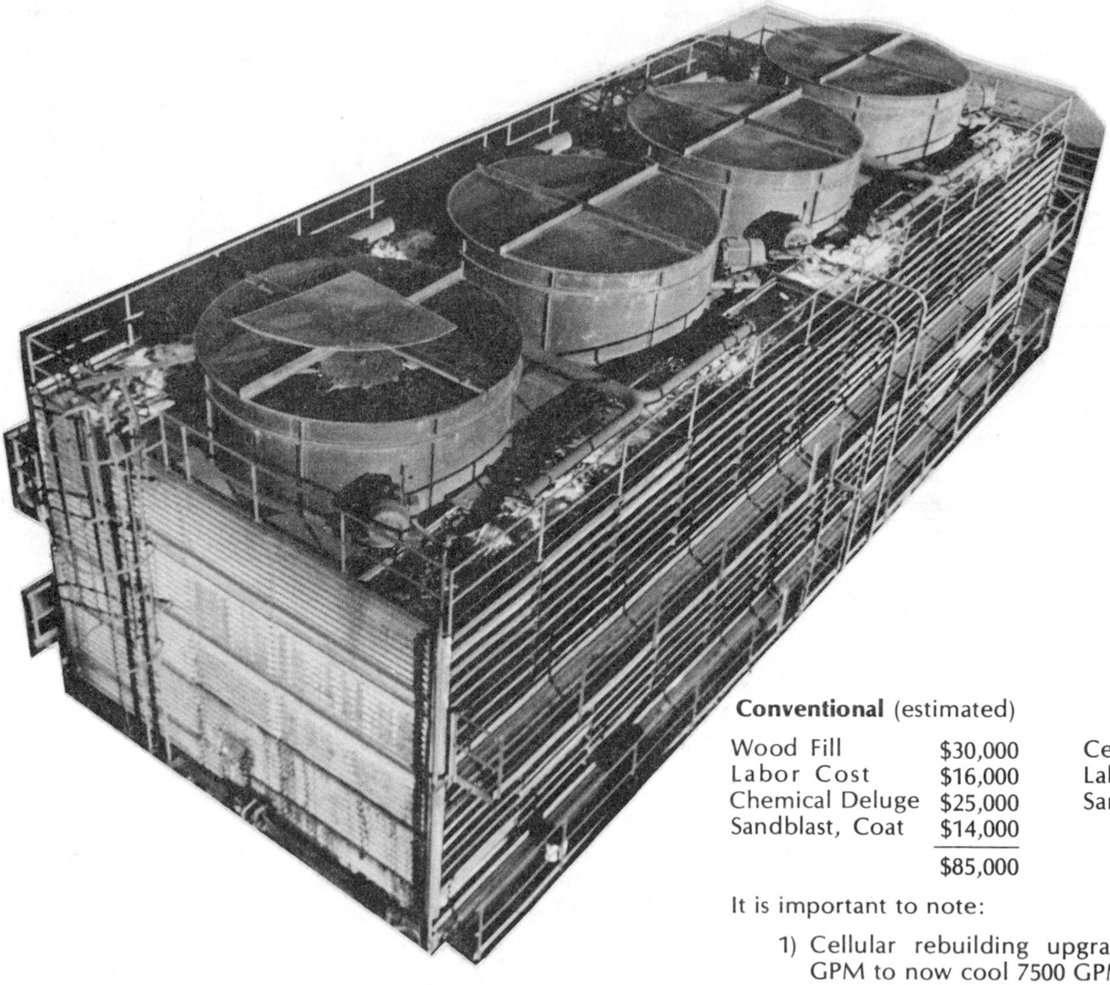

Conventional (estimated)		**Actual Conversion**	
Wood Fill	$30,000	Cellular Fill	$43,000
Labor Cost	$16,000	Labor Cost	$11,000
Chemical Deluge	$25,000	Sandblast, Coat	$14,000
Sandblast, Coat	$14,000		
	$85,000		$68,000

It is important to note:

1) Cellular rebuilding upgraded the original 6300 GPM to now cool 7500 GPM to Design.
2) Conventional rebuilding would not obtain the specified 1200 additional GPM cooling requirement.
3) To erect an additional 7500 GPM tower in this location, including electrical, plumbing, demolition and rigging cost, an additional $50,000.00 would be required.

Here again, the savings of not running an additional 40 HP motor 12 hours a day, 6 days a week for 9 months in the New York City area adds up to $6,630.28 and 887.84 KWH which, projected on the estimated 10 year life of the conversion, more than pay for the rebuilding costs. Add to this the soaring increases in energy over the next decade. Since the cellular fill is 100% fire-proof, the $25,000 cost of a chemical deluge system is not necessary.

Chemical Plant Process Water

In this example of a Chemical Plant located in Texas (Fig. 14), where the cooling tower was originally designed to cool 800 GPM of water entering the tower at 111° and leaving at 90°, at a 75° wet bulb, illustrates how a change in the wet decking heat transfer surfaces alone produced startling results was illustrated.

Fig. 14: Chemical plant tower which produced $27,800.00 savings in rebuilding by merely changing fill and drift eliminators.

The water was not being returned cold enough to do the work properly, and the production planning unit budgeted $40,000.00 to install another cooling tower to make up the deficiency plus the capability for a 50% plant expansion contemplated in the future. Thermal calculations indicated that with the present water distribution system, 24" of cellular fill would more than adequately return the 800 GPM back to the process equipment at 90°. After conversion, this actually occurred.

Further calculations indicated that the addition of another 12" or total of 36" thick cellular fill, together with a new water distribution system consisting of PVC piping and ceramic square spray nozzles, would be sufficient to expand the water cooling capability of the tower by 50%. Since the cost of this conversion is $19,500.00, we can calculate an actual savings of $20,500.00 plus at least $7,300.00 by not having to operate an additional 25 HP motor for the now non-existent "new" tower, in one year's operation.

Conclusion

A blanket statement cannot be made that "all cooling towers can be rebuilt to save quantities of money, power; or can return water significantly cooled. Each installation must be treated on an individual basis, with thermal, hydraulic, and aerodynamic calculations of existing conditions studied to see where areas of improvement can be largely anticipated. Since considerable sums of money can be disbursed on large size cooling towers, it is strongly recommended that a testing requirement be made part of the contract.

The Cooling Tower Institute has formulated ATC 105 which contains rigid testing procedures for determining whether or not a cooling tower will produce what the customer has been promised. The CTI will contract to have testing performed for the customer at a fee; or a qualified cooling tower rebuilding firm could conduct a test under strict supervision of the owner's representative.

The dollar expense of retro-fitting and rebuilding to upgrade the capacity in obtaining colder water or the capability of putting additional gallonage of the existing system without capital expenditure for a new OEM tower has a rapid pay-back if the proper investigative engineering is performed, provided the work is done competently by a firm with a proven record of success. With proper maintenance and application of sound modern engineering principals, a cooling tower can be given a new lease on life and produce colder water which will assist in putting money in the bank for the owner. ■

LaGuardia Airport
Port of N.Y. N.J. Authority

Original Design Conditions

Cool 9800 GPM of water
Range of 18 degrees from 103°F to 85°F
Approach of 7 degrees to a 78°F wet bulb.

New Design Conditions

Cool 12,250 GPM of water
Range of 17.5 degrees from 102.5°F to 85°F
Approach of 6 degrees to a 79°F wet bulb.

DC-10 taking off on Runway 31 into prevailing wind.
VR Stacks throw exhaust air clear of recirculation area.

Low profile of old Fan stacks causes hot moist air recirculation

Solution

a. Install 14′ high Velocity Recovery (VR) Fan cylinder to prevent recirculation of hot air.
b. Replace flammable wood and drift eliminators with efficient fireproof cellular fill of engineered depth for necessary heat transfer.

Results

a. Rebuilt tower passed certified test with 18% excess cold water margin.
b. V.R. Fan stacks threw hot moist air clear of recirculation area.
c. Efficiency total of cellular fill improvement plus venturi VR stacks efficiency calculated to 1200 kilowatt hours of power saved approximating $8,500.00 at New York City 1973 power rates.
d. Installation of fireproofed fill and eliminators.

Sage Intercontinental Defense Missile Command
U.S. Air Force Base

Photo Credit— Air Force Defense Command

Original Bid Requirements
Install

New Wood Fill Decks
New Wood Eliminators
Galvanized Steel Piping
384 Small Bronze Nozzles
Perform misc. Carpentry and Repairs.

Alternate Bid Recommended
Install

ASBESdek Cellular Fill
Cellular Eliminators Stainless Bound
PVC Water Distribution System
144 Ceramic Spray Nozzles
Perform misc. Carpentry and Repairs.

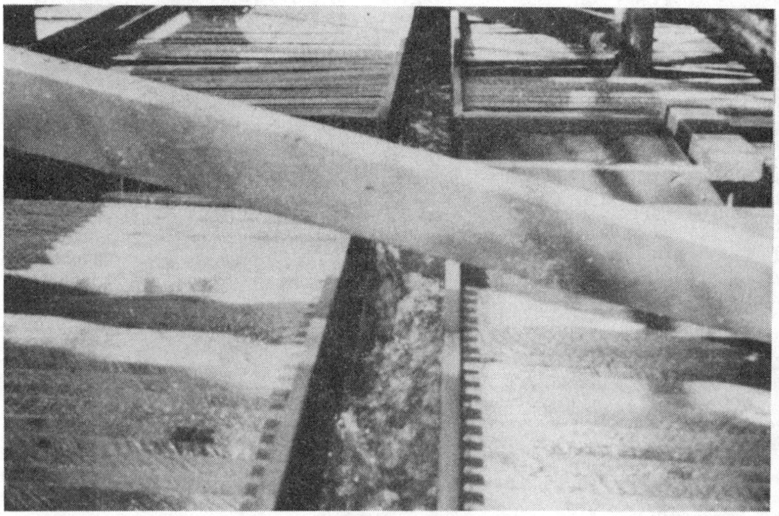

Old watertrough distribution and wood slat drift eliminators.

Installation of PVC piping.

Stainless steel bound eliminators.

Bid Results

Contractor*:	A:	$96,902.00
	B:	$80,000.00
	C:	$73,900.00
	BURGER:	$48,600.00

*Names available on request.

Savings

BURGER ENGINEERED SYSTEM saved the Air Force $25,300.00 over the second bidder and was $48,302.00 less than the high offerer. The Contract was awarded to Robert Burger Associates, and upon completion performed 32% over previous year's operation of cooling 5000 GPM from entering tower at 95°F leaving 85°F at 75°F wet bulb.

John F. Kennedy International Airport
Port of N.Y. N.J. Authority

Original Design Conditions

Cool 35,800 GPM of water
Range of 17.4 degrees from 102.4°F to 85°F
Approach of 6 degrees to a 79°F wet bulb.

New Design Requirements

Cool 52,000 GPM of water
Range of 19.5 degrees from 103.5°F to 84°F
Approach of 5 degrees to a 79°F wet bulb.

13 cell complex

PVC piping and ceramic nozzles on 3' centers

Solution

a. Install cellular fill
b. Engineer PVC water distribution system
c. Increase HP from 100 to 150 per cell
d. Provide 22 foot diameter fans vs 18 foot diameter
e. Use fireproof cellular drift eliminators
f. Add velocity regain fan stacks

Results

a. Additional 52,000 GPM capability for expanded refrigeration and airconditioning requirements provided at an estimated savings over OEM installation of approximately $305,000.00
b. Installation of fireproof fill and eliminators.

Hospital for Special Surgery
Bloomfield, N.J.

Situation

a. Galvanized Steel Fill Eliminators and Piping rotted.
b. Site inaccessible due to Building expansion.
c. Mechanical Equipment and Shell repairable.

Solution

a. Strip interior and mechanically clean.
b. Zinc prime, line with Chlorinated Rubber.
c. Install PVC piping, ceramic nozzles, cellular fill and eliminators.

Before: Package tower apparently headed for scrap pile.

After: Rebuilt with a five-year materials guarantee.

Results

a. Demolition, including removing a wall, rigging and installation of new unit would have cost Owner $12,500.00 (1973).
b. ENGINEERED REBUILDING charge was $3,200.00 resulting in a $9,300.00 savings for the Owner.

KEEP COOLING TOWERS UP TO SNUFF

Inefficiently operating towers waste energy and money. Contractors can assure savings for their customers and profit for themselves by providing proper maintenance and effective upgrading of these vital — but often neglected — system components.

Are all cooling towers on systems you service operating with maximum efficiency?

Not likely!

Those that aren't are wasting energy. And wasting energy means wasting money. Lots of it.

This situation presents you with a golden opportunity for plus business and plus profit. You can point out to your customers how much their inefficiently operating towers are costing them. Then you can offer to service or rebuild these towers, as needed, to eliminate this waste.

If you have the manpower and the know-how, you can do this work within your own organization. If not, you can sub it to a cooling tower specialist. Either way, both you and your customer stand to benefit.

You're familiar with the refrigeration cycle, so you know that the function of a cooling tower is to cool the condenser water after it has removed heat from the refrigerant.

If you know enough about thermodynamics to read a pressure-enthalpy diagram, you can easily figure out that if the water is returned off the tower just 1 degree (F) colder, it will reduce the energy requirements of the system by 3¼%.

Because of their very nature and function, cooling towers often are mounted in places which are very difficult to reach (Fig. 1). This makes it all the more likely that they won't receive adequate maintenance, and will steadily become less efficient. So it's not at all uncommon to find towers in which the slippage from peak efficiency amounts to as much as 3 degrees.

This means that if you can drop that return water temperature back down 3 degrees, you can achieve a nearly 10% reduction in power required to operate the system.

Depending on circumstances, a 1500-ton system can use as much as $300,000 a year in steam or electricity. Saving 10% of that means saving $30,000 a year. And that's not peanuts for any customer.

If the tower water is used for industrial processing, the savings can be even greater. In a chemical plant, for example, it is a relatively simple engineering exercise to determine the effectiveness of colder water on the condensation process.

FIGURE 1 – Atop this skyscraper may be a great place for a high-visibility street number but it's a bad place to have to service a cooling tower. That's why towers like this tend to deteriorate.

A 1-degree drop in water temperature might result in increased production which would net the plant an additional $500 to $1000 per day. This increased revenue can be applied against the cost of upgrading and rebuilding the tower to determine the exact payback period.

Testimonials from users who have had their towers rebuilt confirm the benefits. Here are some typical excerpts:

"Since you rebuilt the tower, our steam consumption has dropped 30% below last season. The drift eliminators not only prevent water loss, but also for the first time the amperage on the fan motors is not exceeding the rating because of excessive back pressure."

". . . Analysis of the first year's operation of the 4-cell, double-flow, crossflow cooling tower more than exceeds your engineering estimates. The 5400 gpm of condenser water is now approximately 2 degrees colder than last year. Head pressures are running 5 to 6 psi lower, which reduces our electrical consumption. One refrigeration machine is now being used 25% less than in previous seasons."

". . . The cooling towers for our hospital airconditioning system never used to produce enough cold water, causing us to operate the system at full capacity. This resulted in high head pressures and excessive electrical consumption. Your recent conversion of these towers, rated at 1200 tons, to cool 3820 gpm of water from 103 F to 85 F, at 78 F ambient wet bulb, has really done the job. The conversion should pay for itself in about 18 months."

Of course, routine maintenance like proper water treatment and cleaning the louvers and drain pans can help. But the real savings and the real profits, as the quoted testimonials suggest, lie in upgrading or rebuilding towers, taking advantage of improved materials and proven engineering principles.

Any competent mechanic can handle the routine maintenance, along with routine or emergency service, but the upgrading or rebuilding requires considerable professional skills. If you don't have these skills, you'd better call in a cooling tower specialist who does, and benefit from his experience.

Upgrading Package Cooling Towers

FIG. 1 — Workman shown during installation of 30" of efficient cellular fill to produce 3 °F colder water.

■ The primary function of a cooling tower is to produce cold water. In too many cases, equipment operates considerably below specified design conditions and does not produce enough cold water for requirements for any of several reasons:

— Higher heat loads were put on since installation;
— Slippage or deterioration occurs as time passes;
— Improper or insufficient maintenance is provided;
— Tower design may initially have been insufficient.

In order to upgrade capacity of a deficient tower, an understanding of the functions of basic components and their relationship is essential to obtain more work from an existing installation. Wet decking fill, water distribution systems, drift eliminators and air handling equipment are prime components which m... be considered.

Two typical upgrading situations applying modern engineering principles and state-of-the-art cooling tower technology follow.

Case 1 — This 10-cell forced draft, blow-through tower (Fig. 3) services equipment that in 1977 generated $450,000 in electrical charges for operating the refrigeration system. As the chart indicates, by changing the internals — steel corrugated wet decking fill to cellular configuration, 3492 small 1/8" diameter orifice brass nozzles to 120 large 1-1/4" diameter nozzles (Fig. 3) and new cellular drift eliminators — an extra 3300 GPM was put on the tower and cooled 4°F lower. This 4°F colder water will save 12-1/2% of the energy cost, amounting to about $56,250. Calculating increasing utility rates, prolonged equipment life and reduced maintenance expenditure, the $95,740 rebuilding cost can be recovered in 18 months.

Case 2 — Engineering analysis of the installation (Fig. 1) indicated that by merely changing the deteriorated fill from the four-year-old Cool-Core to new high heat transfer plastic cellular fill, the $8,740 cost is calculated to be recoverable in six months, resulting from electrical savings due to the 3°F colder water.

FIG 2 — One of 10 cells rebuilt has 12 nonclog ceramic nozzles which deliver 99 GPM each for total of 1188 GPM per cell.

	CASE I		CASE II	
	BEFORE	AFTER	BEFORE	AFTER
Water flow GPM	9500	11,868	3600	3600
Temperature In-Out	95-88	95-84	105-88	105-85
Wet bulb deg F	75	75	78	78
Distribution Nozzle	3492	120	unchanged	
Orifice diameter	1/8"	1-1/4"	unchanged	
Fill	42" steel	36" cellular	42" Cool-Core	30" ASBESdek

FIG. 3 — Three cells of 10 stripped of drift eliminators, water distribution (including 388 nozzles per cell) and wet decking steel plates preparatory to corrosion control procedures and rebuilding to produce 4 °F colder water.

Cold Water Makes Profits

$41,250 Return in One Year On $33,850 Tower Investment

COLDER WATER to refrigeration equipment can reduce power requirements and provide approximately a one-year payback for cooling tower rebuilding.

■The paramount energy conservation factor as it relates to the operation of a cooling tower is not primarily the cost of powering the air drivers. A $10,000 to $15,000 electric bill to run a total of 75 hp in a multiple fan unit for a 1000 ton refrigeration system is not unreasonable. A 15% savings can be accomplished by operating a well-maintained cooling tower. This can amount to a respectable $2000 saving, which is very commendable and professional.

However, the really impressive energy conservation return would be to take a 15% "bite" out of the $250,000 to $350,000 power bill (depending on who sells you the energy) to operate a 1000-ton steam or electrically driven refrigeration unit.

Following is a case history to illustrate this point.

Actual operation — Tower water averaging 88°F; $275,000 in electricity purchased; machines continually on high heads; both 900-ton machines always in use.

Design conditions — Cool 5400 gallons per minute (gpm); leaving machine at 95°F; entering equipment at 85°F; ambient wet bulb 78°F; servicing 1800 tons of refrigeration (Fig. 1).

THE JOB: Recommendations — The tower was inspected by a field engineer, who recommended that the old wood splash bar fill slats (Fig. 1) be changed to highly efficient cullular wet decking heat transfer surfaces. Since the budget was limited, only cells 1 and 3 were converted, and all of the wood was taken out of the tower so that slats in the best physical condition could be put back into cells 2 and 4.

FIG. 1 — Inefficient, deteriorated wood packed splash bars were not producing adequate heat transfer. Replacement was recommended.

PAYBACK AFTER UPGRADING — The building manager and operating engineer reported that the rebuilt cooling tower with cellular wet decking fill in two of four cells returned water 2°F colder (86°F average) and that head pressures were 5 to 6 psi less. One machine which previously had been required to operate continuously was turned off 25% of the time. They estimated that adjusted power costs for running the refrigeration system amounted to $233,750. A savings of $41,250 was returned on an investment of $33,850 in less than one year. Cold water does earn money.

MONEY MAKING SUPERSTAR

One building owner, in one year—
Saved $18,100.00 in fuel oil costs,
Saved $12,650.00 in electrical energy,
Saved $7,380.00 in steam consumption,
Operated compressors at 325 RPM lower speed, and
Ran at 4.45 psi lower head pressures.

All this was accomplished by upgrading his air conditioning cooling tower

12,000 GPM cooling tower with 4 cells at 60 HP each rebuilt so that only 3 cells were required to cool the water to design conditions.

- **Upgraded tower provides 25% standby capacity**
- **Permits 25% additional life for equipment**

Request FREE Cooling Tower Operating and Rebuilding Guide

Elements of Modernized Cooling Tower
A. Fireproof Cellular Fill **C. PVC Non-Corroding Piping**
B. Efficient Drift Eliminators **D. Non-Clogging Ceramic Nozzles**

Energy Conservation Strategies For Cooling Towers

Cooling towers can play a major role in the upgrade, performance and cost efficiency of a facility's energy program

Because of their seeming simplicity, cooling towers are often the step-children of the process plant, power generating equipment, and refrigeration system. While engineers are convinced of the importance of sophisticated plant equipment, and rightly so, they take the cooling towers and the cold water returning from them for granted.

Design conditions for the particular requirements are specified before a cooling tower is purchased. After it is put on line and the cold water temperature or volume becomes inadequate, plant personnel look to solutions other than the obvious. While all cooling towers are purchased to function at 100% of capability in accordance with the required Design Conditions, in actual use, the level of operation may be considerably different.

A number of factors affect cooling tower demand including:

•Present service needed is greater than the original requirements.

•Slippage because of use or deficient maintenance has reduced performance of the tower.

•Installation could have been originally undersized because of the "low bidder syndrome."

•New plant expansion needs additional water volume and probably colder temperatures off the tower.

Tower Upgrading

Users of cooling towers are not particulary concerned with the thermal analysis involving calculus, or delving into the complicated hydraulics and aerodynamic principles of equipment selection. The plant engineer is primarily concerned with maximizing equipment performance.

Three elements of all counterflow and crossflow cooling towers should be investigated when upgrading a cooling tower to produce higher levels of cooler water, which conserves energy in the plant. They are the air handling system, water distribution, and heat transfer surfaces.

Air Handling

The least expensive fix to improve air handling and increase the volume is by pitching blades to the maximum angle consistent with the motor plate amperage requirements.

The installation of velocity regain (VR) fan cyclinders will generate approximately 7% more air because they relieve the exit pressure which the fan works against. This brings into effect Bernoulli's principle, which states that increasing the velocity of a fluid reduces its pressure, or conversely reducing the pressure of a fluid increases it's velocity and effective volume. Installing a larger motor and/or fan should be considered to increase air flow and obtain higher performance levels.

Water Distribution

Crossflow water distribution patterns are set by the flow requirements and fixed position orifices (holes) located in the hot water distribution basin. The conventional orifice drops the solid column of water on a lattice work of wood which splashes the water about the top of the fill. A higher level of performance can be obtained by removing the splash deck and replacing it with efficient redistribution decks or target orifice nozzles. Both procedures provide a more uniform distribution at the top of the tower, and use the entire height for cooling, rather than a portion of it for water breakup, and a lower net temperature results.

Existing spray systems are greatly improved by installing non-corroding Polyvinylchloride (PVC) piping in conjunction with nonclogging, non-corroding square spray ABS plastic nozzles.

Heat Transfer Surfaces

The most dramatic performance improvement is obtained by installing, on a retrofit basis, cellular fill. Whether it be counterflow or crossflow configurations, change to a high-efficiency, dense-film fill, also known as cellular fill, results in improvements of as much as 5° to 10° F colder water, which could equal up to more than a 50% increase in capacity of the existing tower. The main factor in this improvement is that conventional splash bar towers have water bouncing from slat to slat on it's way from top to bottom; only the exterior of the droplet is cooled, (Figure 1). If a high enough vertical travel is engineered, the water will be cooled to Design Conditions. However, the more wood in the tower, the higher the static pressure loss and larger horsepower motor and fan diameters are required.

Cellular fill works on the principle that the droplet of water is stretched into a thin film as the water proceeds vertically downward through the cells and the entire droplet is cooled more rapidly (Figure 2). Even though the static pressure per cubic foot of cellular fill is much higher than the static pressure per cubic foot of wood fill, considerably less depth of cellular fill is required to produce the same or greater cooling results; therefore less total pressure drop occurs, which provides a greater available volume of usable cooling air.

While the fill upgrading changeout from wood splash bars to cellular could be the most expensive first cost retrofit procedure, it produces the highest level of improvement in performance which will provide for a more rapid Return on Investment (ROI).

Function Of Lower Temperatures

The role of the cooling tower is to remove waste heat. In a condensation process, colder condensing water temperatures increase unit production at a lower unit cost. The degree of elevation of the process temperature above ambient conditions is the sum of the tower's approach of the cold water to the Wet Bulb temperature, the cooling range (which equals the temperature rise in the heat exchanger), and the terminal difference in the exchanger. A reduction in operating temperature, always desirable for economic reasons, is obtained by increasing the capability of the cooling tower's performance.

The importance of colder water for the compression of gases is evident in that all compressors have one act in common: about 80% of the energy is converted to heat, which must be continuously removed at the same rate it is generated or the compressor will overheat and shut down. Reducing the operating temperature of the compressor proportionately reduces the energy input requirements. In other words, the colder the water returns to the equipment, the less energy is required to produce the same degree of work at lower costs.

In a refrigeration or cooling system, colder water requires less energy for operation. Enthalpy charts indicate that for every 1° F of colder water returning from the cooling tower within the operating range of the compressors, approximately 2% less energy will be required by the compressor to produce the same cooling results. Thus, a cooling system consuming $350,000 of electric or steam energy to power the turbines can reduce this utility cost by about $32,000 per year for a 4° F colder water from the tower.

Counterflow Tower

The importance in value of the utilization of colder water to plant engineers is clearly indicated from a well-engineered and rebuilt cooling tower functioning at optimum levels of performance.

When a plant manager is told by management that the production capacity of his plant or the refrigeration requirements are going to be increased, or for example, doubled, he is faced with a decision of installing a new cooling tower or investigating the retrofit possibilities of upgrading the existing equipment.

In this case history, Design Conditions of the two-cell counterflow tower were

FIGURE 1

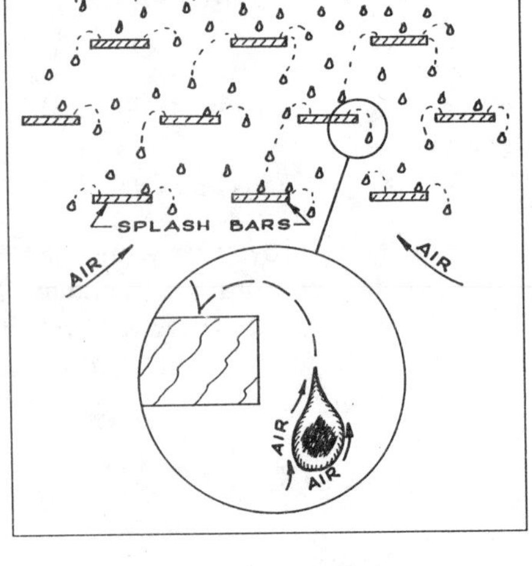

FIGURE 2

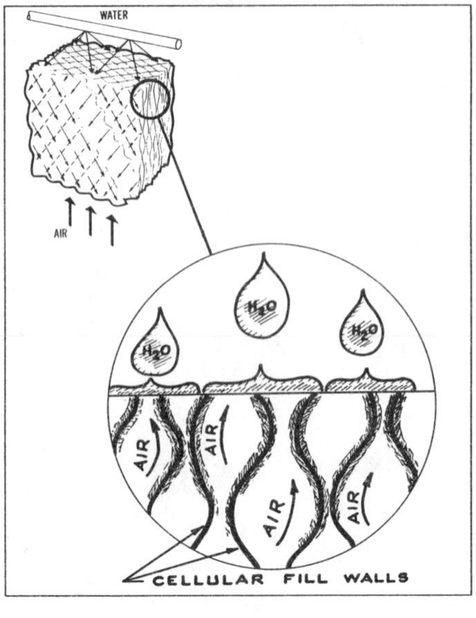

Installing "V" bar fill in crossflow towers can either produce colder temperatures or cool additional water volume to original design.

originally to cool 6,500 GPM from entering the tower at 108° F, leaving at 90° F, during an ambient Wet Bulb temperature of 78° F. The manufacturer's data and associated fan curves indicated that the two 18-foot diameter air foil aluminum fans utilizing 40 horsepower drivers would produce 300,000 actual cubic feet per minute (ACFM). This was designed to provide a cooling range (Delta T) of 18° F with a 12° F approach to the Wet Bulb.

To yield the projected quantity of increased product, 10,500 GPM of water, approximately 60% greater capability of the existing 6,500 GPM rate must be circulated through the system, which initially indicated erection of a new tower.

However, new tower delivery scheduling, plot plan constraint problems and budgetary requirements dictated that if the existing tower could be rebuilt to accommodate the newer capacity, considerable savings in time, real estate, and dollars would accrue over the alternative installation of new OEM equipment.

In order to provide sufficient cooling water for the doubled output of the new projected requirements, the old 6,500 GPM tower would have to be upgraded to cool 10,500 GPM of water from entering at 108° F at a 78° F Wet Bulb.

It was also evident that if the additional gallonage was not added to the tower after retrofit, the rebuilt facility would have the capability of cooling the 6,500 GPM from 108° F to 84° F at a 78° F Wet Bulb indicating a new capacity of providing a 24° F Delta T with a 6° F approach to the Wet Bulb which would be 6° F colder than before upgrading.

A further discussion at management level indicated that a new requirement of doubling the plant's output again was being considered which would then require cooling 15,500 GPM of cooling water. Engineering calculations indicated that with additional rebuilding, the existing tower would have the capability of cooling this increased requirement.

This clearly illustrates the viability of modern retrofit techniques which can upgrade the original design from cooling 6,500 GPM to a new level of 15,500 GPM.

The upgrading and modern retrofit consisted of changing the water distribution from old-fashioned water troughs and inefficient wood splash bar wet decking fill to high-efficiency heat transfer Polyvinylchloride (PVC) cellular fill and a PVC non-corroding water distribution system utilizing large orifice 1 1/4-inch diameter ABS nonclogging square spray pattern nozzles.

The installation of the high thermal transfer, cellular fill surfaces provided over a 50% increase in capability, new water distribution patterns were extremely uniform, and the new cellular drift eliminators functioned at lower static pressure levels.

Crossflow Tower

Due to its cost per cubic foot, cellular fill is restricted to about 15 feet in height for industrial plant cooling towers. Larger towers can be retrofitted economically to produce colder water by installing "V" bars which are a combination of splash and film cooling.

A Midwest manufacturing plant had to increase the circulating water volume from 20,000 GPM to 28,000 GPM. This was easily accomplished by taking out the old style wood splash bars and engineering a "V" bar installation in the 20-year-old double flow, crossflow cooling tower.

Conclusion

The cooling towers are hidden bonanzas for energy conservation and dollar savings when properly engineered and maintained. In many cases, the limiting factor of production is the quality and quantity of cold water coming off the cooling tower. The savings accrued in energy conservation and additional product manufactured can be an important factor on the operator's company's profit and loss sheet.

Energy management analysis is a very important consideration in today's escalating climate of energy costs. It is advisable to consider a thorough engineering inspection and evaluation of the entire plant to leave no stone unturned in the search to reduce energy consumption. The cooling tower plays the major role in heat waste removal and should be given a thorough engineering inspection and evaluation by a specialist in this field. This can be performed at nominal cost and a formal report submitted with recommendations, budget costs, and evaluation of the thermal, structural, and mechanical condition of the equipment. This feasibility study will assist in determining the extent of efficiency improvement available with costs and projected savings.

It can be stated that practically all cooling towers can be upgraded to perform at higher levels of efficiency which can provide a rapid, cost-effective payback. However, while all cooling tower systems might not provide such a dramatic cost payback as these case histories, the return of a customer's investment in upgrading his cooling tower can be a surprising factor of operation and should not be neglected.

Robert Burger is president of Burger Associates, Inc., Consultants, engineers and contractors specializing in cooling towers. The company is a corporate member of AIPE.

Square spray pattern on cellular fill is today's state-of-the-art upgrading.

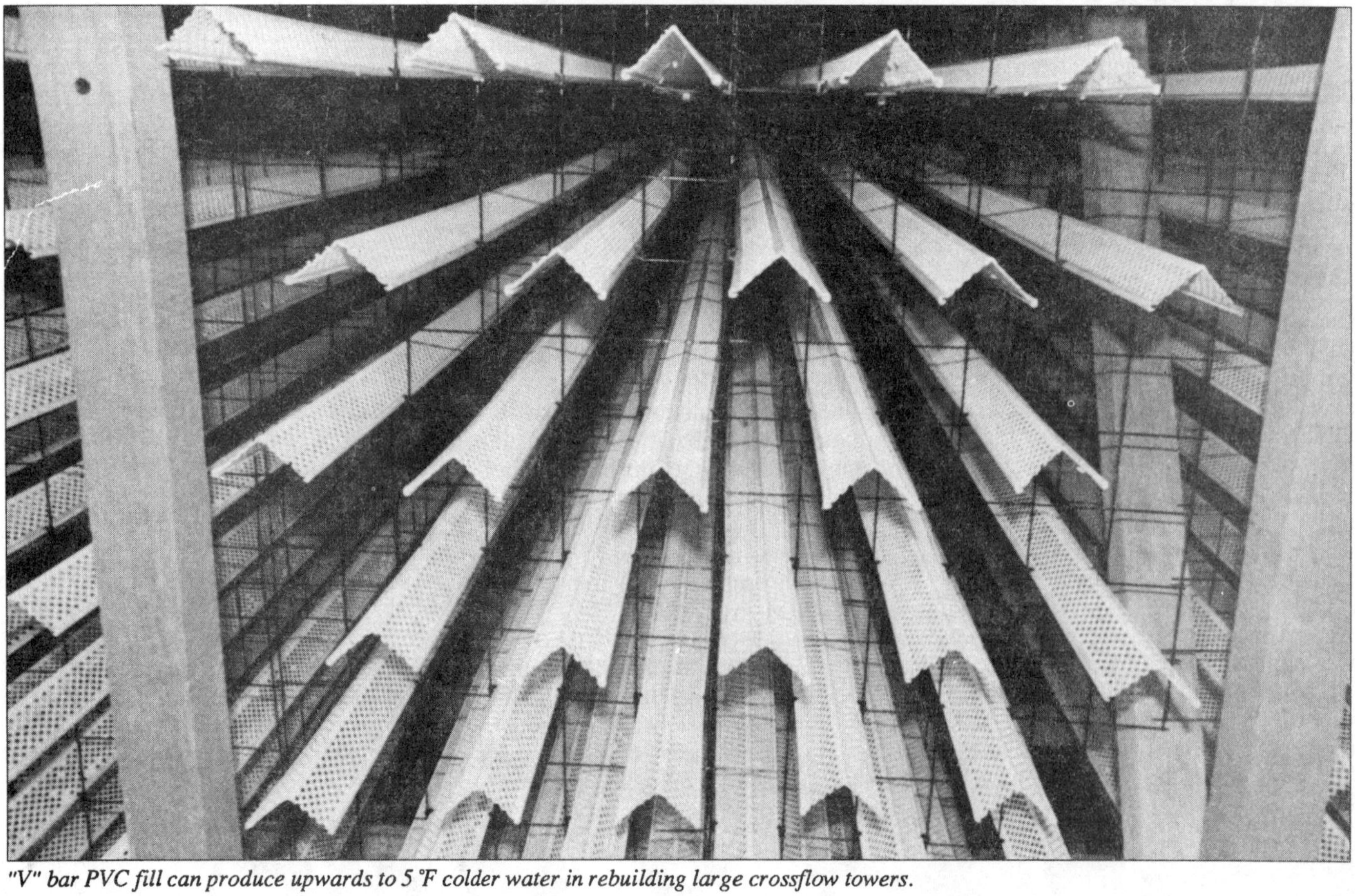

"V" bar PVC fill can produce upwards to 5 °F colder water in rebuilding large crossflow towers.

Retrofitting cuts cooling-tower costs

Faster cost payback can follow from upgrading a cooling tower than from expanding or replacing it. Here are some ideas on how towers can be revamped to produce larger quantities of colder water.

The role of the cooling tower is, of course, to remove waste heat. In a condensation process, it is obvious that colder condensing water will increase unit production and reduce unit cost. The degree of elevation of the process temperature above normal operating conditions depends on the tower's approach of the cold-water to the wet-bulb temperature, the cooling range (which equals the temperature rise in the heat exchanger), and the terminal difference in the exchanger. A reduction in operating temperature, always desirable for economic reasons, may be obtained by increasing cooling-tower performance [*1*].

The importance of colder water in the compression of gases is evident from the fact that most of the energy (80%) to all compressors is converted to heat. This heat must be removed at the same rate at which it is generated or the compressor will overheat and shut down. Reducing the operating temperature of a compressor will proportionately reduce the energy it consumes [*2*].

A refrigeration system, whether for process or comfort cooling, operates on less energy if the water is colder. Enthalpy charts indicate that for every 1°F of colder water (within the operating range of the compressors), 3-1/3% less energy is required by a compressor to produce the same cooling [*3*]. Thus, 4°F colder water from a cooling tower can cut over $46,000 from a $350,000 annual cost of electric or steam energy to power the turbines.

Virtually all cooling towers can be upgraded to perform at a higher level of performance with a rapid payback of the investment involved. There are three major elements of the cooling tower generally worth investigating with a view to upgrading: air-handling arrangement, water-distribution system, and heat-transfer surfaces.

Air handling

The least expensive way to increase air volume through a forced-draft tower is to pitch the fan blades at the maximum angle permitted by the motor plate amperage.

Another way to boost air volume is to install velocity-regain fan cylinders, which will generate approximately 7% more air movement, with no increase in motor horsepower, by reducing the exit pressure against which the fan works (Fig. 1). (This follows Bernoulli's theorem, which states that increasing the velocity of a fluid reduces its pressure; or, conversely, reducing the pressure of a fluid increases its velocity, and thus the effective volume.) Replacing a right-angle fan deck and fan cylinder with a velocity-regain stack will streamline the air inlet and increase air movement.

Of course, air flow can also be enlarged by installing a larger motor or fan, or both.

Water distribution

Crossflow patterns of water distribution are set by the flow requirements and the location of orifices, which are fixed in the hot-water distribution basin. The conventional orifice drops a solid column of water on a lattice work of wood that splashes the water about the top of the fill.

Distribution can be improved by removing the splash deck and replacing it with efficient cellular redistribution decks or target orifice nozzles. Both will provide a more uniform distribution at the top of the tower, thereby using the entire height of the tower for cooling rather than having a part of it functioning only to break up and splash the water.

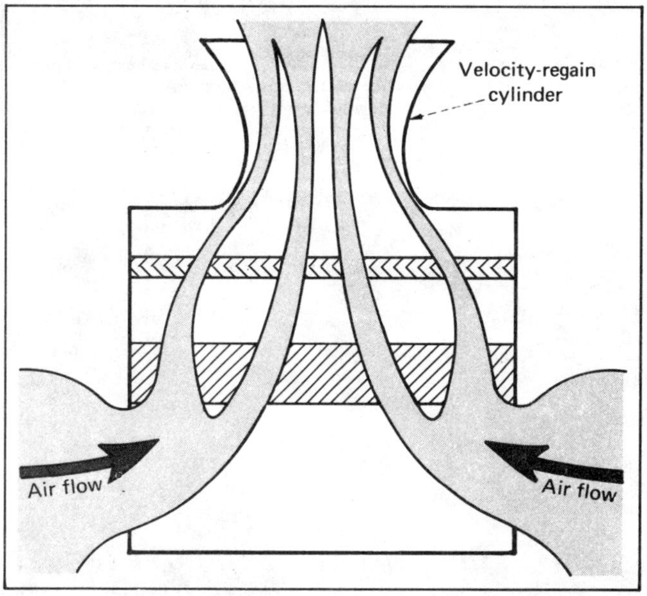

Figure 1—Venturi effect of velocity-regain cylinder reduces exit pressure, boosting air movement by 7% without a hike in motor hp

Water troughs or enclosed flumes in counterflow towers should be changed to a low-pressure spray piping system. Existing spray systems can be greatly improved by installing noncorroding polyvinyl-chloride (PVC) piping, together with nonclogging, noncorroding square-spray acrylonitrile-butadiene-styrene (ABS) plastic nozzles.

Heat-transfer surfaces

Retrofitting a tower with cellular fill can dramatically improve its performance. Whether the configuration is counterflow or crossflow, this dense-film fill can reduce the water temperature by as much as 10 to 15°F, which could increase the capacity of some towers by as much as 50%.

In a tower, crossflow or counterflow, with conventional splash bars, the exterior of droplets is cooled as the water bounces from slat to slat on its way down (Fig. 2). If this vertical travel is long enough, the water will be cooled to design conditions. However, the more wood there is in the tower, the higher the pressure drop and the larger the motor horsepower and fan diameter required.

The cellular fill stretches the droplet into a thin film as the water passes down through the cells, allowing the crossflow or counterflow air to cool the water more rapidly (Fig. 3). Although the pressure drop per cubic foot of cellular fill is much higher than that with wood fill, considerably less depth of cellular fill is required to produce the same or greater cooling results, thereby generating less total pressure drop and providing a greater volume of usable cooling air.

Although retrofitting a tower with cellular fill is usually very expensive, the improvement in performance will provide a rapid payback.

Retrofitting in an ammonia plant

Output of anhydrous ammonia was to be increased from 120,000 to 225,000 tons/yr. The cooling tower structure, fabricated of old-growth clear-heart redwood, was about 25 years old. It contained eight cells, each with a 20-ft-dia. fan driven by a 75-hp (56 kw) motor. It was designed to cool 36,000 gal/min of water (4,500 gal/min per cell) entering at 115°F and leaving at 85°F, with the wet-bulb temperature at 78°F. This produced a 30°F cooling range with a 7°F approach.

Because the tower was overdesigned initially, doubling anhydrous ammonia output only required that cooling-water circulation be increased by 6,600 gal/min to 42,600 gal/min.

It was strongly recommended that a ninth cell be added (crossflow configuration with the fan powered by a 100-hp motor). However, the project engineer decided that the matter should be investigated more closely.

Figure 2 — With the conventional wooden splash-bar slats, only the exterior surface of a water droplet is cooled

Figure 4 — Square-spray nozzles and cellular fill can together significantly boost water capacity and lower water outlet temperature

Figure 3 — Cellular fill stretches water droplet into a thin film that is cooled faster and colder by air at lower operating pressure

Figure 5 — Drift eliminators of PVC will retain water droplets to less than legal minimums and help keep tower pressure drop at low level

An inspection revealed that the tower's structure, connectors and mechanical equipment were in excellent condition. Thermal calculations indicated that — with the installation of 24 in. of cellular fill at the bottom of the tower (above the air-intake louvers) — the existing eight cells could, with no additional horsepower, easily cool 42,600 gal/min of water (5,325 gal/min per cell).

Furthermore, the new fill could be installed in such a way as to allow additional fill to be added at a later date to again upgrade the tower's capacity. To improve water distribution, the 25-year-old spray-nozzle system was replaced by non-clogging ABS nozzles (Fig. 4). Because the original polystyrene fill was in relatively good condition, a large expenditure for labor was avoided by not removing it. Instead, the existing wet-decking fill was augmented by the addition of a 24-in. layer of high-efficiency PVC-film fill.

The savings from rebuilding the existing eight cells instead of adding a new cell was about $100,000. However, of greater significance was the ongoing savings from the fact that 100 hp (75 kW) was not required to operate a ninth cell. With electricity at 7¢/kWh, operating a fan motor for 8,760 h/yr would cost $55,869/yr. Over the projected 10-year life of the tower, taking into account estimated escalations in the cost of utilities, more than $1 million in operating expense would be saved. Thus, upgrading the tower was clearly more economical than adding to it.

Figure 6 — V-bar fill, a combination splash- and film-cooling configuration, will produce colder water than will conventional wooden slats

Table — Annual costs for running 10 to 200-hp fan motors

Motor size, hp	Full-load kVA at 1,800 rpm	Operating cost, $/yr (8,760 h) at $0.01 kWh*
10	10.14	888.3
15	15.67	1,372.7
20	20.06	1,757.3
25	24.93	2,183.9
30	28.58	2,503.6
40	38.32	3,356.8
50	47.10	4,126.0
60	57.21	5,011.6
75	71.10	6,228.4
100	91.11	7,981.2
125	118.47	10,378.0
150	140.57	12,313.9
200	188.65	16,525.7

*To determine motor operating cost at a rate other than $0.01 kWh, multiply figure in third column by the particular rate multiple; for example, if the cost of electricity is $0.07 kWh, the annual cost of operating a 100-hp motor = $7,981.2 x 7 = $55,868/yr.

Retrofitting in a power plant

The original operating conditions of a 35-year-old double-flow crossflow cooling tower in a power plant were as follows: 35,000 gal/min capacity, 114°F inlet water, 91°F discharge water, and 80°F wet-bulb temperature. The tower had six 24-ft-dia. fans, each run by a 75-hp motor. To expand the power-generating capacity of the plant, the following tower operating conditions were required: 42,000 gal/min of water entering at 114°F and discharging at 88°F. The existing conventional wooden-slat wet-decking splash-bar fill was in moderately sound condition, but the heat-transfer rate was considerably below that required of the new design.

The drift eliminators, which were inefficient as well as broken, allowed an excessive quantity of water droplets to be drawn out of the tower. These were replaced with high-performance PVC drift eliminators (Fig. 5). Thermal calculations indicated that, if the efficient V-bar fill (Fig. 6) were installed on 8-in. vertical centers and 4-in. horizontal centers, the splash and film surfaces provided by this configuration would adequately meet the new design requirements.

Because the mechanical-equipment rails had deteriorated, entire units had to be removed by crane, rebuilt, then returned into position. To handle the additional volume of water, the distribution holes in the hot-water basin were enlarged and orifice target nozzles of PVC were installed.

Adding a new cell would have required extending the concrete basin as well as the plumbing and electrical wiring. The engineering, fabrication and erection would have taken four months at a turnkey construction cost of $310,000. In addition, a new cell would have required a 75-hp motor, which would operate 24 h/d for 48-50 wk/yr at a cost of $30,000/yr, with electricity at 5¢/kWh (see table).

Rebuilding was selected because, in addition to the savings, the tower had to be put back on line after a 14-day plant turnaround. This required having two crews work 10-hour shifts. The fill, drift eliminators and equipment rails were all retrofitted within schedule at a cost of $181,740 — a capital cost saving of $128,260, to which can be added the operating cost saving of $30,000/yr.

J.Matley, Editor

References

1. "ASHRAE Equipment Handbook," American Soc. of Heating, Refrigerating and Air Conditioning Engineers, 1791 Tullie Circle, NE, Atlanta, GA 30329, 1975, Chap. 21, p. 11.
2. Stoven, K., Energy Evaluation in Compressed Air Systems, *Plant Engineering*, June 1979.
3. "The Pressure Enthalpy Diagram: Its Construction, Use, and Value," Allied-Signal Corp., P.O. Box 1057R, Morristown, NJ 07960.

Modernizing Crossflow Cooling Towers To Bring Equipment Into The 1980's

Figure 1 — Older crossflow cooling towers have tremendous potential for upgrading their performance through modernization.

Energy considerations and budgetary restrictions dictate that a prudent analysis be made of rehabilitating existing equipment to thermally upgrade it for energy optimization and durability consistent with long range planning.

Twenty years ago, cooling towers (Figure 1) were constructed of old growth Clear Heart Redwood. These structures are usually solid today. The drawback however, to these towers is that the thermal heat transfer is using a 25- to 35-year-old technique. Whereas using today's state-of-the-art, discharge water can be produced 4 to 10°F colder.

A typical engineering investigation consists of four areas of concern:

1. Structure.
2. Heat Transfer.
3. Drift Eliminators.
4. Mechanical Equipment.

These elements will be discussed using modern reconstruction techniques with before and after situations.

STRUCTURE

Typical areas of wood rot attack are the joists, beams, and tops of vertical columns supporting the fan deck and mechanical equipment supports. (Figure 2) Particular attention is paid to rusted connectors (once galvanized) which have caused debilitating iron rot that has attacked the wood at the bearing surfaces.

Figure 2 — Top decking joists usually show signs of wood rot and can be readily changed out to new redwood beams.

Figure 4 — Old fashion thin section wood splash bar slats have poor heat transfer capability.

The mechanical equipment ranges from 75 to 200 HP and 18 foot to 32 foot diameter fans. It is very important to renew all of these supporting members where required, (Figure 3) since the starting torque and rotational forces could result in physical damage and an expensive outage. Many fan decks are fabricated from plywood or other relatively thin wood, however the extra expense of utilizing 6″ x 2″ tongue and groove Redwood will pay for itself in longevity. A thorough inspection of all interior columns and braces should be made while the fill is out of the tower and individual structural replacements can easily be effected.

Figure 3 — The fan deck and mechanical equipment supports are renewed using chemically treated California redwood.

Figure 6 — "V" bar combination splash and film self extinguishing PVC blades produce colder water for the same given air flow and water circulation.

Figure 7 — Inefficient wood drift eliminator blades have high static resistance and permit solid droplets of water to escape tower.

Figure 8 — Modern PVC wave form drift eliminators assist in producing colder water with minimum drift.

HEAT TRANSFER

Typical heat transfer surfaces on the old style towers are narrow, ¼″ thick wood splash bar slats. (Figure 4)

This wood is relatively thin and the California Redwood Institute states that the service life of thin section Redwood used in cooling towers is approximately 15 to 20 years. Therefore, in these older-type towers the fill is usually deteriorated to the point where a change out would be required to maintain any semblance of heat transfer.

A very professional analysis of the performance of many configurations of crossflow fill is investigated and calculated by Neil W. Kelley in his publication "Kelley's Handbook of Crossflow Cooling Tower Performance." This handbook is the accepted standard in the cooling tower industry for cross-flow analysis.

This publication provides heat transfer curves and methods of predicting the performance of various materials. It also includes illustrations and the procedures for calculating and projecting performance values of various types of fill configurations.

Figure 5 is typical, wherein, the performance of wood slats (Figure 4) is compared to "V" bars (Figure 6). The PVC self-extinguishing "V" bars, a combination splash and film cooling surface, are highly efficient. In the subject cooling tower of this article they have produced 6 °F colder water after conversion.

To convert wood to the more efficient PVC, the old existing fill must be taken out and disposed of from the premises, along with their hanging supports. New supports to fit the configuration of the "V" bars are then installed. The new materials are properly "threaded in" as indicated in Figure 6.

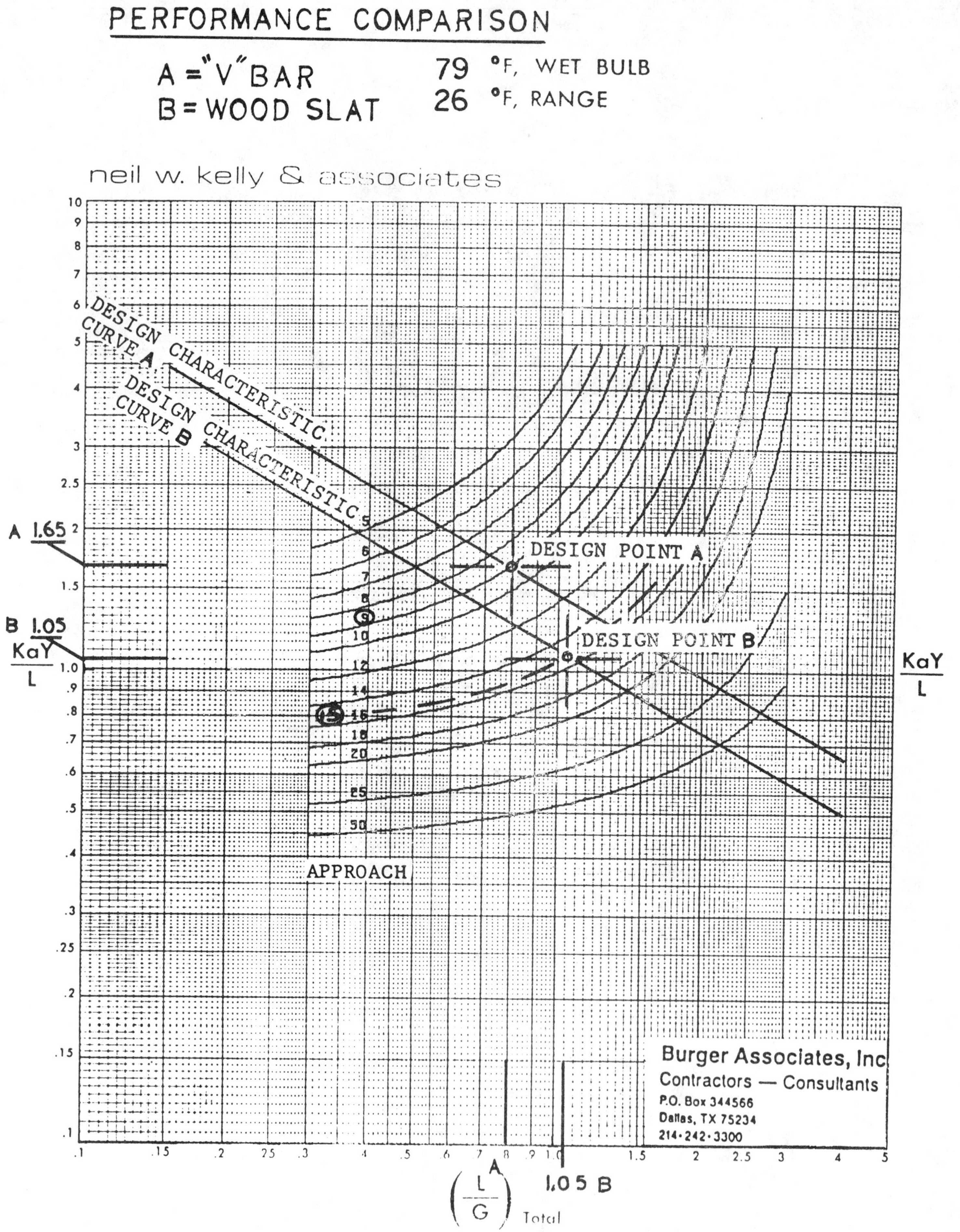

Figure 5 — Kelley's performance curves indicate that "V" bar fill will produce a calculated 6°F closer approach to the wet bulb.

DRIFT ELIMINATORS

A 2 pass "herringbone" wood bladed eliminator section is typical of the old towers antiquated construction. These units deteriorate, along with the thin section fill slats, and create leaks, loss of water. They also expose equipment adjacent to the cooling tower to a rust conducive environment.

The modern drift eliminator replacement consists of wave form PVC as in Figure 8. In addition to being at least 50% more effective than the wood slat eliminators, the specially engineered and designed blades have a lower pressure drop. This permits a more effective use of the airflow in cooling the water.

MECHANICAL EQUIPMENT

Repair or replacement of the mechanical equipment should be considered at this point, since inadequate air movement due to under sized motors or fans will seriously affect the performance of the tower. Equipment of this size and weight is best removed from the tower and lowered to the ground where it can be easily worked on. In the cooling tower being reviewed, the flexible shaft couplings were renewed, new input gearbox oil seals on all 4 cells were installed, and the motors were sent to a motor shop for overhaul and maintenance work. After re-assembly the units were then put back into position on the new supporting members. (Figure 9)

Thermal engineering calculations can be readily worked out for the entire system if required for upgrading purposes. Fan manufacturers, for example, provide curves of air fan performance. Using these, the thermal engineer can determine the adequacy of the existing equipment to meet the new design requirements of the power plant or chemical plant facility. As a comparison, the chart below indicates the before and after performance of this 28-year-old doubleflow crossflow cooling tower, together with the reconstruction cost and estimated comparison of new OEM equipment.

Figure 9 — Mechanical equipment re-installed after repairs are affected while they are out of the tower.

CONCLUSION

The budgets show that it is obvious, on both an engineering and financial basis, to thoroughly investigate the rebuilding of an existing large cooling tower before considering it obsolete. One important factor to mention is that a rebuilding can be performed during a "turn around" period whereas a new tower requires 6 to 8 months of planning. This would probably include a new concrete basin, new piping, and revision of the electrical installation.

The rehabilitation work described in this article was performed during a 20 day scheduled outage. All materials were delivered to the job site prior to starting date, a special completion time bonus was offered to the motor shop (an offer they couldn't refuse) and two superintendents worked a crew of 25 men to bring the turn-key job in on time.

Unless customized and built directly to the purchasers requirements, most new towers, do not contain the higher quality materials that are used in rebuilding due to the competitive nature of the market. All stainless steel connectors (nuts, bolts, washers, ring barb nails, TECO rings) are used in reconstruction while new towers commonly use galvanized fittings.

A professional evaluation can determine the dollar and time savings accrued and calculations of cold water return will have an important bearing for the budgeted ROI. □

Design Specifications
Four 26 foot diameter fans
Four 100 horsepower motors

Original Design		New Performances
30,000	GPM Water	36,000
114	Entering °F	114
91	Leaving °F	88
79	Wet Bulb °F	79
23	Range	26
12	Approach	9

Rebuilding Cost $375,000.00
Estimated OEM Cost $600,000.00

Retrofit your cooling towers to save energy and money

Properly engineered and maintained cooling towers can save energy and dollars

ENERGY CONSERVATION strategies in cooling towers involve more than a cursory examination of fan horsepower consumption. The optimization of cold water and analysis of pumping head can provide vast savings potential. Two important questions must be answered:

1. What is the dollar value of 1°F colder water in your process? and,
2. What is the dollar cost of pumping water to the distribution system?

Colder water economics. Cooling towers are very much maligned and neglected pieces of equipment. Cooling towers are usually ignored until a crisis occurs and then a quick fix is performed that may or may not be the best to optimize the effectiveness of the equipment.

Retrofit counterflow. The basic contract for the five-cell counterflow cooling tower (Fig. 1) was to upgrade it to the original design conditions that it was never able to produce, of: cooling 15,000 gpm (58,000 ℓpm) of circulating water when entering at 120°F (58.8°C) and leaving at 85°F (29.4°C) during a 75°F (23.8°C) ambient wet bulb temperature.[1]

Prior thermal testing of the tower indicated that during full operation at design water loading and design wet bulb, the cold water was returning at 89.6°F (32°C). This 14.6°F (9.6°C) approach to the wet bulb, as against the 10°F (3.3°C) required, was causing high heads and wasting energy.

The inefficient wood splash bars and deteriorating water distribution system (Fig. 2) were the main causes of the tower's heat transfer deficiency.

Thermal calculations[2] indicated that with a 36-in. (0.91-m) installation of high-efficiency cellular fill, in conjunction with a properly engineered and designed water distribution system, using PVC piping and ABS noncorroding square-spray pattern (Fig. 3), the design conditions could be met.

The cost of this retrofit was $97,000 and since the refrigerant enthalpy diagram indicated that for each 1°F (0.5°C) colder water, 3.5% less energy[3] would be required, and the Return on Investment (ROI) would be less than two years. On a ten-year depreciated basis, assuming that the 8¢/kWh-

Fig. 1—Five-cell cooling tower with obsolete octagon configuration of three cells and 46-ft-high riser pipes.

Fig. 2—Original inefficient wood splash bar and deteriorating water distribution system.

TABLE 1—Refrigerant enthalpy diagram

1°F Colder water
= 3.5% less energy required to operate compressors

4°F Colder water × 3.5% = 14% less input energy needed

14% × $350,000 = $49,000 per year
or
$490,000 + earned for 10 years

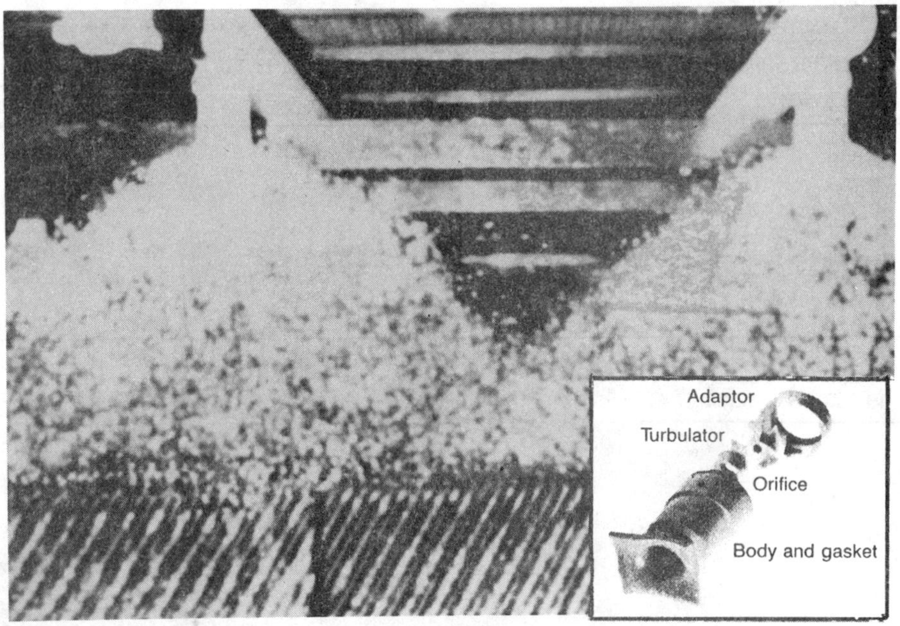

Fig. 3—Retrofit of noncorroding PVC piping and ABS square-spray-pattern nozzles with high-efficiency cellular fill gives dramatic results.

cost in the area will rise, the ten-year savings is over $500,000 for a $97,000 investment (Table 1).

Pumping head economics. To further optimize the cooling tower's utility, and after discussions with the owner's engineers, it was decided to maximize the energy conservation by reducing the pumping head of the tower. The rationale for this was based on the super-efficient cellular fill's maximum requirements of 36 in. (0.91 m) height against the original 30 ft (9.1 m) height of conventional wood splash bars.

In addition to reducing the pumping head from 42 ft (12.5 m) and 46 ft (14 m) respectively to the new 22 ft (6.7 m) height, the V sections of the older three cell towers were eliminated to provide interior heat-transfer surface area.

The thinking behind the octagon design configuration was that there would be less turbulence (static pressure loss) of the exit air through the cooling tower vs a square or rectangular box. While this could be true to a certain degree, the extra volume available for heat transfer more than makes up the possible incremental loss of static pressure. In any event, the octagon shape has lost favor with design engineers and has not been manufactured for at least 20 years.

Fig. 4 shows the overall appearance of the old configuration. The two cells (4 and 5) have been structurally altered, the new roof deck and safety rail have been reinstalled prior to reducing the riser pipe's height. The existing elbow and flange is then re-welded to the riser piping at the new 22-ft-high pumping head.

After completion (Fig. 5) the tower was tested according to the Cooling Tower Institute Acceptance Test Code (ATC-105) and indicated 96.4% of the new design capability, which was within the contract requirements of 95% to 100% of capability (Table 2).

Data on the rebuilt cells justify the expenditure of the $84,675 extra to the base contract. The five cells circulate 3,000 gpm (22,375 ℓpm) or a total of 15,000 gpm (58,800 ℓpm). To determine BHP savings, the energy costs of Table 3 were used.[4]

The hydraulic calculations (Table 4) indicate that 94.7 hp can be eliminated over the original pumping requirements if the cooling tower is operated for the maximum available time during the 8,760 hour-year. In the BHP formula, the pumping head reduction should be the average of the three cells at 42 ft (12.5 m) and two cells at 46 ft (14 m) but needless to say, an average of 20 ft (6.1 m) calculated savings in pumping head indicates (Table 3) that over $63,000 or approximately 800,000 kWh was conserved. In ten years, the savings can exceed $750,000 or 8,000,000 kWh.

While utilization of the cooling tower complex varies, the ROI will be quite rapid, since the initial-upgrading savings can be added to the electrical pumping head savings:

	Cost	Savings	
		One year	Ten years
Thermal upgrade	$ 97,400	$ 49,000	$ 500,000
Pumping head reduction	84,675	63,000	750,000
Total	$182,075	$112,000	$1,250,000

Crossflow upgrading. The decision to rebuild existing large-capacity double-flow crossflow cooling towers for chemical plants or industrial power stations to provide necessary additional capacity for plant expansion, or to augment the performance with a new OEM erected cell requires inspection and evaluation to select the most economically viable strategy.

Many power plant installations and petrochemical plants require expansion necessitating additional water volume through the cooling tower and/or colder water discharge to maximize production and meet new requirements. Many times the existing cooler tower lacked capacity in the beginning.

The original operating conditions and the new design requirements for the field-erected double-flow crossflow cool-

TABLE 2—Record summary sheet*

Tower design conditions:			
Tower water U.S. gpm	15,000	Inlet wet bulb, °F	75
Elevation, ft	SL	Design L/G	1.07
Fan hp/cell	50	Design ACFM	388,000
No. of cells	5	Range	35°F
Ka V/L	1.95	Approach	10°F
Hot water temp., °F	120	Pumping head	44 ft
Cold water temp., °F	85	Tds, PPM	N R
Tower Test Conditions:			
Water flow U.S. gpm	14,346	Cold water temp., °F	87.8
Cells tested	5	Inlet wet bulb, °F	70.7
Fan hp/cell	39.1	Test L/G	1.56
Pumping head	22 ft	Range	33.1°F
Ka V/L		Approach	17.1°F
Hot water temp., °F	120.9	% Capability	96.4

* Corrected for make-up water and heat added by the pump if the cold water temperature is taken downstream of the pump.

TABLE 3—Electric motor operating costs

Motor size hp	Full load kVA @ 1,800 rpm	Operating cost per year (8,760 h) at stated cost per kWh ($) .03	.04	.05	.06	.07	.08	.09	.10
10	10.14	$2,665	$3,553	$4,441	$5,330	$ 6,218	$ 7,106	$ 7,994	$ 8,883
15	15.67	4,118	5,491	6,863	8,236	9,609	10,982	12,354	13,727
20	20.06	5,271	7,029	8,786	10,544	12,301	14,058	15,815	17,573
25	24.93	6,551	8,735	10,919	13,103	15,287	17,471	19,655	21,839
30	28.58	7,511	10,014	12,518	15,022	17,525	20,029	22,532	25,036
40	38.32	10,070	13,427	16,784	20,141	23,498	26,855	30,211	33,568
50	47.10	12,378	16,504	20,630	24,756	28,882	33,008	37,134	41,260
60	57.21	15,035	20,046	25,058	30,070	35,081	40,093	45,104	50,116
75	71.10	18,685	24,913	31,142	37,370	43,599	49,827	56,055	62,284
100	91.11	23,944	31,925	39,906	47,887	55,869	63,850	71,831	79,812
125	118.47	31,134	41,512	51,890	62,268	72,646	83,024	93,402	103,780
150	140.57	36,942	49,256	61,570	73,884	86,198	98,511	110,825	123,139
200	188.65	49,577	66,103	82,629	99,154	115,680	132,206	148,732	165,257

TABLE 4—Hydraulic calculations

$$\text{BHP} = \frac{\text{gpm} \times \text{pumping head}}{3{,}960 \times \text{efficiency} \times \text{sp. grav.}}$$

$$= \frac{15{,}000 \times (42' - 22')}{3960 \times 0.80 \times 1.0}$$

$$\text{BHP} = 94.7 \text{ saved}$$

$$20 \text{ hp @ 8¢/kWh} = \$14{,}058$$
$$75/95 \text{ hp @ 8¢/kWh} = \underline{49{,}827}$$
$$= \$63{,}885$$

Fig. 4—During reconstruction, cells 4 and 5 have been structurally altered prior to reducing pumping heights.

ing tower in this example (Fig. 6) are:

Old operating conditions	New design requirements
Three 28 ft dia. fans	Existing mechanical equipment
Three 125 hp motors	36,000 gpm
30,000 gpm	114°F entering temp.
114°F entering water	88°F discharge temp.
91°F discharge water	80°F wet bulb temp.
80°F wet bulb temp.	

The old conventional wood-slat wet-decking splash bar fill was in moderately sound condition, but the heat transfer rate was considerably substandard of design requirements and the tower never performed at design. The drift eliminators were inefficient, broken-up and permitted excessive solid-water droplets to be drawn out of the tower. Fig. 7 shows the new PVC high performance drift eliminators that were retrofitted. This new concept in cooling tower drift eliminators features lower pressure drop and can operate at higher air velocities with higher separation efficiency than conventional blade-type drift eliminators. This type of installation can reduce the drift to below 0.008% of the circulating water volume.

Thermal calculations indicate that if the highly efficient PVC fill bar (Fig. 8) was installed, the splash and film surface provided by this configuration would adequately meet the new design requirements.[5] To handle additional water volume the distribution holes in the hot water basin were enlarged and metered orifice PVC target nozzles were installed.

Cost comparisons. To engineer, fabricate and erect an additional cell would require a new concrete basin extension, piping, electrical and four months to erect. The turnkey construction cost would be $310,000. In addition, the new cell would require a 125-hp motor operating 48 to 50 weeks per year, 24 hours per day, which would require approximately $72,464 in electricity at an average cost of 5¢/kWh (Table 3). This electric penalty was added and escalated yearly for a 10-year life.

Rebuilding was selected, because in addition to the considerable savings, the new facility had to be put on-line after a 14-day plant turnaround. This required two crews working

Fig. 5—Completely rebuilt cooling tower tested at 96.4% of design.

Fig. 6—Three-cell, 30,000-gpm tower upgraded to 36,000 gpm with no increase of input horsepower.

10-hour shifts to meet the tight schedule. The new self-extinguishing PVC fill, drift eliminators and mechanical equipment were all retrofitted within the tight time-frame. The work was performed at a cost of $181,740 or a savings of

Fig. 7—High-performance cellular drift eliminators installed.

Fig. 8—Mechanical equipment modernization also increases the capability of a cooling tower.

$128,260 over the new cell option for the utility company[6] less the add-on cost of $76,646 per year (Table 5); i.e., Cell No. 4 and a 125-hp motor were not required. On the basis of ten years of operation at a constant utility charge of 5¢/kWh, adds to $894,740 (Table 3), or on a projected electrical-charge escalation, over $1,000,000 would be saved for a $181,740 investment. These data are shown in Table 5. With normal plant maintenance the tower should have a minimum of 10 to 15 years additional life.

In many cases cooling towers are the limiting factor of production due to the quantity and quality of cold water leaving the tower. The savings accrued in energy conservation and additional production can be important factors on the profit and loss sheet.[7]

Energy management analysis is a very important consideration in today's escalating energy costs. A professional engineering inspection and evaluation of the entire plant should be made to reduce energy consumption.[8] The cooling tower should be given a thorough engineering inspection and evaluation. This can be performed at nominal cost and a report submitted with recommendations, budget requirements and an evaluation of the thermal, structural and mechanical condition of the equipment. This feasbility study will assist in determining the extent of efficiency improvement available with costs and projected savings. Practically all cooling towers can be upgraded to perform at higher levels of efficiency, which can provide a rapid cost-effective ROI.

TABLE 5—Rebuilding economics versus new cell

OEM Cell 4 quote	$ 310,000
+ 125 hp motor/year	72,646
	382,646
Rebuilding cost	− 181,740
Initial savings	200,906
9 yr 125 hp @5¢/kWh	894,740
10 year savings	$1,095,646

LITERATURE CITED

1 East Coast nuclear plant, name and location upon request.
2 Burger, R., Cooling tower technology," Chapter 8, (1979).
3 Allied Chemical Corp. "The pressure enthalpy diagram: It s construction, use and value."
4 Hunt-Atwater, K. M., BAC-Pritchard, CTI paper, 1982.
5 "Kelleys handbook of crossflow cooling tower performance," 1976.
6 Florida Public Utility, name and location upon request.
7 Willa, J., "Proper cooling tower operations make money," Cooling Tower Institute Annual Meeting, 1975.
8 Mathur, S., "Energy Today," *Hydrocarbon Processing*, July 1980.

APPENDIX 1

GLOSSARY - THE LANGUAGE OF COOLING TOWERS

the language of cooling towers

Technicians, engineers, and operating personnel using sophisticated mechanical equipment, develop a jargon over the years. On some occasions, this private language relegates information only to the "members of the club."

Purpose of this *glossary on cooling tower* terms is to disseminate information, since everyone at some time or another is a victim of technological one-upsmanship. How many times has a person used terms and expressions to you, and you nod your head wisely hoping that he will clarify what he is talking about so that your lack of knowledge on the subject matter will not be apparent?

This glossary is to assist in better understanding of cooling tower operation. While thermo-dynamic heat transfer theory has been lightly touched, the main thrust concerns everyday operations. For most operating engineers, design technology is a moot question since the equipment is already designed and operating.

ACCEPTANCE TESTING

Uniform standards to determine the water cooling capability of cooling towers. Instrumentation and measurements are performed in accordance with methods outlined in the Cooling Tower Institute (CTI) Bulletin Acceptance Test Procedure (ATP) 105 or ASME PTC 1958. Tests should be utilized to ascertain that new or rebuilt cooling towers produce Design Conditions that the Owner is paying for. Testing normally is done by third party—not the Owner or Contractor. Confidential report is submitted to person paying for testing and the Owner.

ACCESS TUNNEL

Opening in Natural Draft Tower used for access to the cold water basin. It can be large enough for a man only, or large enough for mechanical equipment.

AIR FLOW

Total amount of dry air and associated water vapor flowing through the tower, measured at exhaust from the tower and converted to standard air which has a density of 0.075 lb. per cu. ft. Units: quanitity, lb. per hr; velocity, fpm. Standard air = (acfm) (density at DBT, WBT and pressure/ 0.075).

AIR HORSEPOWER

The measure of useful power required to move a given air rate against a given resistance. The ratio of air horsepower to fan input horsepower is the measure of fan efficiency.

AIR INLET

Area of cooling tower where air is initially introduced into unit by action of air flow generator.

ALGAE

A low form of plant life which generally requires sunlight and air for existence. Causes plugging of heat exchanger tubes and cooling tower distribution systems.

ALGAECIDE

A toxic material which will retard or prevent the growth of algae and slimes. Some of the more commonly used algaecides are chlorine, copper sulfate and phenolic compounds.

AMBIENT TEMPERATURE

External outdoor temperature as reported by periodic readings, also known as the dry bulb temperature, measured in the regular manner with conventional instruments.

AMBIENT WET BULB TEMPERATURE

Wet bulb temperature of air measured windward of the tower and free from the influence of the tower. Unit: °F. °C.

ANCHOR BOLT

A threaded bolt imbedded in a concrete basin or fitted to supporting members, to which an anchor casting is attached.

ANCHOR CASTING (Column Anchor)

A device for attaching the tower structure to the foundation; it does not include the anchor bolt.

APPROACH

Temperature difference in degrees between the water leaving the tower and the wet bulb temperature of the air entering the cooling tower. Performance is measured in terms of APPROACH. The cost of a 6°F approach tower for the same heat load would be approximately 50% higher than a unit cooling to a 10°F approach.

ATMOSPHERIC TOWER

One in which air movement is dependent upon atmospheric conditions, not on mechanical fans, to force or induce air through the tower. The hot moist air rises in the chimney pulling in colder outside air through the intake louvers.

AUTOMATIC VARIABLE PITCH (AVP)

Sensors and pneumatic controllers change fan blade angles to utilize minimum HP depending upon the varying heat rejection load required at any given time. High Energy Saver.

AXIAL FLOW

Air movement parallel to spinning axis generated by rotating fan blades.

CENTRIFUGAL FAN causes air movement at right angles to rotating impeller axis.

BALANCING VALVE

Hand or mechanically operated valve installed in each riser pipe of a multicell tower to permit altering the water flow to the unit to equalize or change the volume over each cell.

BASIN

Bottom collecting area of cold water returning to suction line. Crossflow tower has additional top hot water distribution basin and in some few cases, an intermediate hot water distribution basin is located halfway between the top basin and the bottom collecting basin.

BASIN CURB

Top level of the retaining wall of the cold water basin; usually the datum point from which tower elevation points, tower static and total pumping heads are measured.

BAY

The area between two bents or lines of framing members.

BENT

A line of structural framework composed of columns, girts or ties; a bent may incorporate diagonal bracing members.

BEVEL WASHER

A metal fitting used to accommodate through-bolts to angular position of a diagonal member, usually connecting to columns or other framework members.

BLOWDOWN (Purge)

Remove impurities concentrated in circulating water due to intensification of solids by water evaporation loss, by bleeding off a small percentage of the total flow.

BOARD FOOT

A unit of measurement of lumber represented by a board 1′ long, 12″ wide, and 1″ thick or its cubic equivalent.

BROWN ROT

In wood, decay in which the attack concentrates on the cellulose and associated carbohydrates rather than on the lignin, producing a light to dark brown friable residue.

CANOPY (Natural Draft Tower)

Connects the hyperbolic shell to the cooling section, acting as an air conduit and air seal between the two.

CAPACITY

Average amount of circulating water volume in the cooling system at any given time expressed in terms of gallons per minute (GPM) flow.

CASING

Enclosure for sidewall and endwall of a counterflow tower and endwall of a crossflow tower exclusive of the louvers.

CELL

One complete unit of a cooling tower consisting of one distribution system, normally one set of mechanical equipment and partition walls. Each cell can be designed to operate independently of its neighbors and receive an equal share of the total water volume if required for the flexibility of the system.

CELL DIMENSIONS

(a) Width: dimension perpendicular to tower longitudinal axis and usually at right angles to the louver area; (b) Length: dimension parallel to longitudinal axis and the plane where louvers are usually placed; (c) Height: distance from basin curb to top of fan deck but not including fan stack. Nominal width and length are measured from and to the column centerlines. (see Nominal Tower Dimensions).

CELLULAR FILL

Most efficient fill packing developed. Spreads water droplets into thin sections throughout the cells of the fill thereby cooling a larger surface area for the same energy. A good design will have a lower static pressure—for the same amount of work performed—than wood splash bar packing or steel configuration.

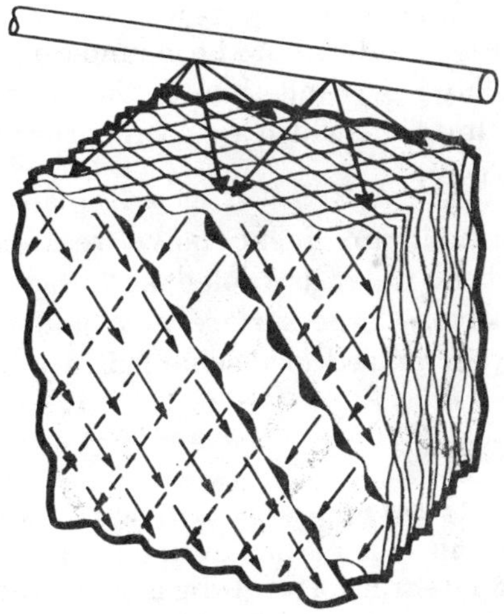

CELLULAR FILL most efficient heat transfer surfaces developed for cooling tower use.

CENTRIFUGAL FAN

Forced draft or blower wheel (also called "squirrel cage") causing air movement at right angles to rotating impeller axis.

CHARACTERISTIC

KaY/L (Crossflow) or KaV/L (Counterflow) is a measurement of the order of difficulty for the cooling tower heat rejection requirements exclusive of the fill. The characteristic is located on the vertical or "Y" scale, the ordinate. The value is expressed as a dimensionless number and is usually obtained by locating the Liquid to Gas ratio and intersection of the approach curve. Integration of the Tchebycheff Formula will also locate the value on the vertical trigonometric scale.

CIRCULATING WATER FLOW

The amount of hot water flowing into the tower. Units: quantity, lb. per hr. (one U.S gallon = 8.33 lbs.); volume, GPM; velocity, FPS.

COIL SHED

Portion of the tower structure housing atmospheric (tubular) heat exchangers.

COLD WATER TEMPERATURE (CWT)

Temperature of the water entering the cold water basin before addition of make-up or removal of blowdown.

COLUMN

Framework member; a main vertical supporting member in the tower framework.

CONVECTION

Transference of heat by the circulation of the heated parts of gas or liquid whereby heated portions are lighter and rise vertically.

COOLING FACTOR

L/G (L over G)—Ratio of the pounds of water being circulated per unit time to the pounds of dry air per unit time which are cooling the water.

COOLING POND

Large reservoir on inexpensive real estate whereby warm water is cooled as air contacts pond's relatively large surface area and colder water is either reused or discharged into public waterway.

COOLING TOWER

A device for the evaporative cooling of water by contact with air. This is achieved partially by an exchange of latent heat resulting from the evaporation of some of the circulating water, and partially by transfer of sensible heat to the air.

COUNTERFLOW

Tower where air movement and hot water mix at 180° counter to each other with the air moving vertically through the fill packing and water.

CROSSFLOW

Tower where air movement and hot water mix at 90° or cross to each other with the air moving horizontally through the fill packing and water.

CROSS OVER PIPING (See Header)

CROSS STRUTS (Natural Draft Tower)

Framework that holds up the hyperbolic shell, the number of which will vary depending on shell size.

CYCLES OF CONCENTRATION

Compares dissolved solids in make-up with solids in the circulating water. Since chlorides are soluble, for example, cycle of concentration is equal to ratio of chlorides in circulating water to chlorides in make-up.

DECIBLE

Unit of sound level measurement, gauged by converting the pressure energy of noise to electrical impulses and transferring them to a voltmeter calibrated in decibles.

DECK STRINGER

Holds the splash bars of the fill deck in a fixed position with respect to air and water flow.

DELIGNIFICATION (Soft Rot)

Loosening of surface fibers of softer lignites causing weakening of timbers or splash bars, reducing members in cross-section and strength.

DESIGN CONDITIONS

Thermal parameters for which the cooling tower is purchased and/or rebuilt. Expressed as cooling a given GPM flow of water entering the tower at a specified temperature, cooling through a given range leaving the tower at the required temperature and having a designated approach to a stated wet bulb. A certified test (see Acceptance Testing) should be incorporated in the job specifications.

DE-SILTING SUMP (Natural Draft Tower)

Area in the cold water basin, usually a low point, where any silt can be flushed to a drain.

DEW POINT

The temperature at which a given mixture of air and water will have a relative humidity of 100% saturation.

DIAGONAL (Brace)

Framework member; any load-bearing member transmitting forces at other than a right angle with reference to columns or horizontal ties.

Crossflow Cooling Tower

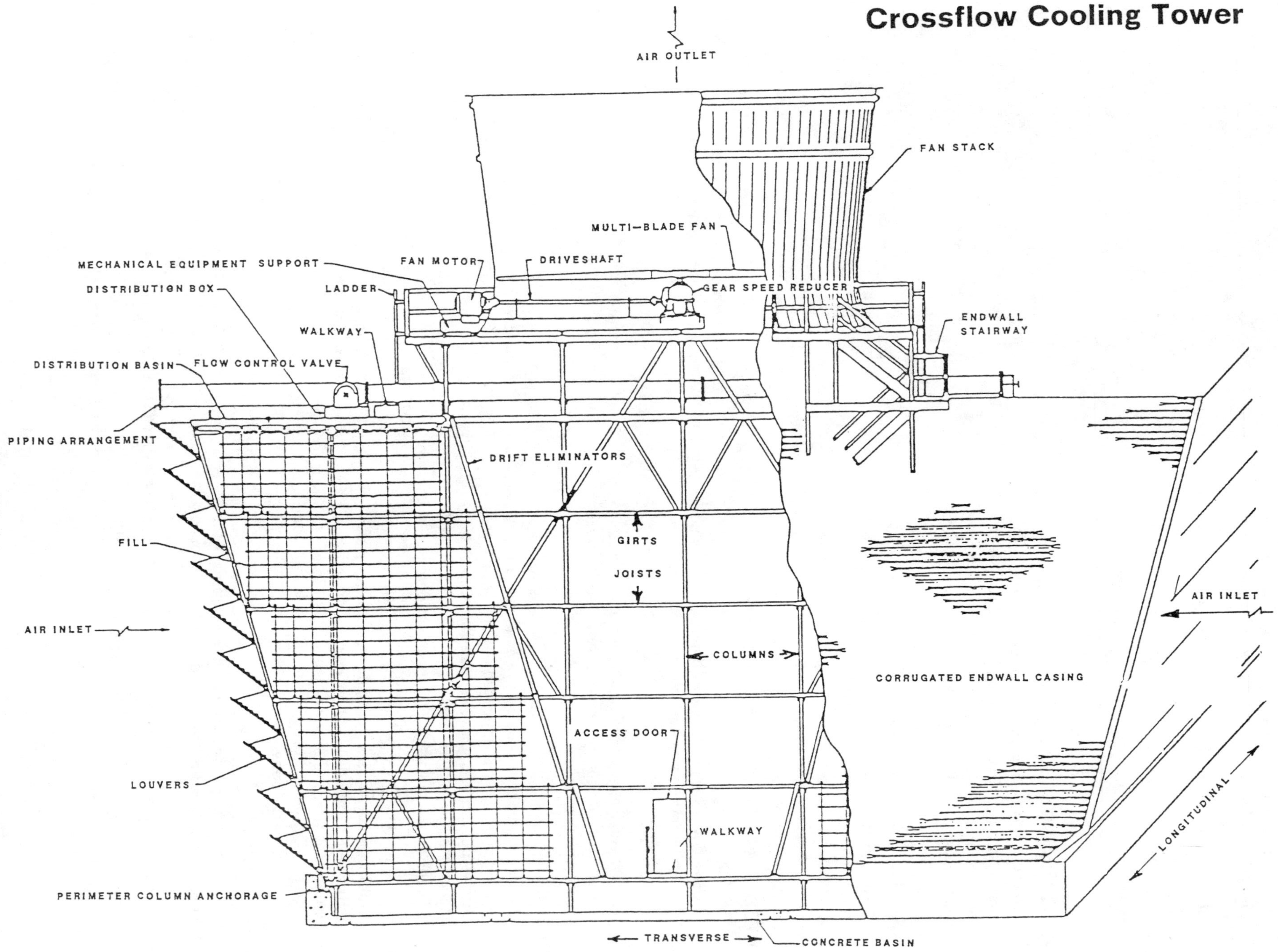

DIFFUSION DECK

A fill deck which is located directly under the distribution basin or nozzle bank. The purpose of this deck is to receive water from the basin or nozzle and distribute it uniformly over the fill decks.

DISCHARGE STACK

A walled enclosure extending upward above the eliminators to direct exhaust air vertically away from fans in a forced draft tower.

DISTRIBUTION BOX

Used in conjunction with the manifold and valve assembly in a Crossflow Tower to disperse the hot water uniformly in all directions thereby increasing the effectiveness of the distribution nozzles.

DISTRIBUTION SYSTEM

Mechanical method of uniformly passing the hot water over the fill area preparatory to cooling. Low pressure spray through piping and nozzles is normally used in counterflow tower, and gravity drop is mainly utilized in crossflow towers.

DOUBLE-FLOW WATER COOLING TOWER

A Crossflow Tower with two fill sections and one plenum chamber which is common to both.

DOWNSPOUT

A short vertically placed pipe or nozzle used in a gravity distribution system to divert water from a flume or lateral to a splasher.

DRIFT

Water droplets which are entrained in the airstream as it passes through the tower and are thrown out of the hot air discharge plenum or fan stack. Manufacturers limitations are 2/10ths of 1% allowable, but vary with specifications.

DRIFT ELIMINATORS

Baffling that causes discharging hot air with entrained water droplets to change direction a number of times, thereby causing the droplets to hit the eliminator surface and fall back into the tower at every change of direction.

DRIVE SHAFT

A device including a coupling (or couplings) for transmitting torque from the driver to the speed reducer.

DRIVER

Primary drive for the fan drive assembly; usually an electric motor.

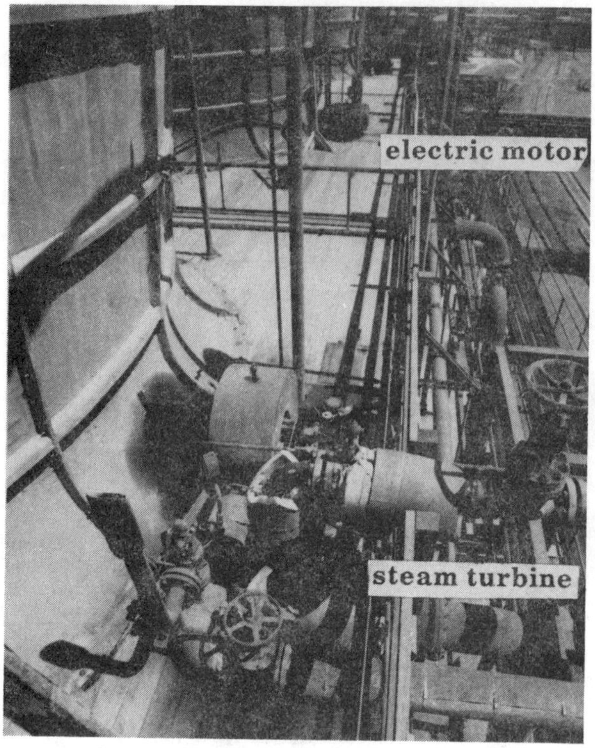

DRIVER primary mover for the fan drive assembly; usually an electric motor, but may be steam turbine or other power source.

DRY BULB

Ambient temperature of air measured in the regular manner with conventional instruments.

DRY ROT

A term loosely applied to any dry, crumbly rot but especially to that which, when in an advanced stage, permits the wood to be crushed easily to a dry powder.

ELIMINATOR BOARD

The smallest component in a wood drift eliminator assembly which is usually installed in a fixed position at an angle to the direction of air flow. Also known as Eliminotor Baffle (Blade).

END WALL

The wall on the end of the tower structure.

ENTHALPY

The total heat content, which is the sum of the sensible heat of the air and water vapor plus the latent heat of vaporization of the water.

EVAPORATION LOSS

Water evaporated from the circulating water into the atmosphere by the cooling process.

EVAPORATIVE COOLING

Heat transfer where a liquid ~~condenses~~ into its gaseous state thereby giving up latent heat. The simplest example of this basic premise of cooling tower operation: wet the back of your hand then blow on the wetted surface. The drop in temperature on your skin is due to evaporative ~~condensation.~~

EXHAUST AIR

The mixture of air and its associated vapor leaving the tower.

EXHAUST WET-BULB TEMPERATURE

Average wet bulb temperature of the air discharged from the tower.

EXIT BASIN TEMPERATURE

Temperature of the circulating water leaving the cold water basin. If blowdown is removed from or make-up added to the basin, the temperature will be affected accordingly.

EXIT DIAMETER (Natural Draft Tower)

Diameter of the shell at the top.

FAN

An air foil rotating to move air through a cooling tower normally parallel to the axis of the fan shaft.

FAN DECK

Surface enclosing the top of an induced draft tower. In a counterflow tower, the fan deck covers the entire top surface of the tower. In a crossflow tower, the fan deck covers only the tower plenum area, leaving the distribution system exposed. An extended fan deck encloses the distribution system and covers the entire top surface of the tower, except of course, the cut for fan diameter.

FAN DRIVE ASSEMBLY

Mechanical components furnishing power to the fan, usually consisting of driver, drive shaft, speed reducer, and supporting members.

FAN DRIVER INPUT

Horsepower input to the driver. For 3-phase alternating current (ac) motors.

$$\text{hp} = (\text{amps})\,(\text{volts})\,(3)\,(\text{Power Factor})\,(\text{Efficiency}) \div 746$$

FAN DRIVER OUTPUT

Brake horsepower output of the driver to the drive shaft. (Fan driver input) $\times$ (motor efficiency).

FAN GUARD

A protective screen installed either at the inlet of a forced draft fan or at the exit of an induced draft fan.

FAN PITCH

The angle which a fan blade makes with the plane of rotation; degrees from horizontal.

FAN STACK

Cylindrical or modified cylindrical structure enclosing the fan discharge in induced draft towers. Usually constructed of fiberglass materials, wood and metal are also utilized.

FILL BARS

The assembly of splash bars comprising the tower filling. Fill bars intercept the downward fall of water at regular intervals, forming splash surfaces which cause water drops to break into smaller droplets, and provide wetted surfaces for air-water contact.

FILL DECK

The assembly of splash bars comprising the tower filling. See FILL BARS for description of operation in a Crossflow Tower.

FILL HANGER

Support system in a Crossflow Tower for fill bars which hold fill in place.

FILL PACKING

Baffling placed within the casing of the tower to provide large water surface areas for heat transference. Two classes of materials are used. Splashbars of wood, metal, transite, plastic, or film pack (cellular fill) made of thin section. The splash type cools water droplets bouncing down in the vertical or horizontal air stream, while film packing turns the droplets into thin films for highly efficient cooling by producing a larger surface area for the air movement to cool.

FLEXIBLE COUPLING

Resilient connecting assembly that transmits the power in the fan driver to the gear unit through the drive shaft.

FLOW CONTROL VALVE

A manually controlled valve generally located in the hot water supply line.

FLUME

A trough which may be totally enclosed or open at the top. Flumes are sometimes used in cooling towers to distribute the hot water over the fill.

FOG

Saturated air being discharged from the cooling tower under atmospheric conditions whereby the ambient atmospheric temperature cannot absorb all of the moisture as vapor, and excess fog condenses. The intensity of the fog plume is a function of the heat rise of air passing through the tower and the temperature and humidity of the ambient air. Fog plumes are normally permissible since there are no droplets of water raining out of the discharge area. However, on large industrial sites this could cause icing of the roads in the winter and restricted visibility which could necessitate corrective action.

FORCED DRAFT

Design whereby the air intake is at the bottom forcing the air up and through the tower, normally by a centrifugal blower.

FOUNDATIONS (Natural Draft Tower)

Support material beneath the tower cross struts and cooling section. Foundations can be caissons, spread footing (tension ring) or piles.

FRAMEWORK MEMBERS

The structural members designed to support all live and dead loads imposed. They consist of columns, horizontal ties, diagonals, and joists and beams.

GEARBOX

Power transmission unit on cooling tower which provides a right angle turn from motor shaft and couplings to gear down motor speed thereby generating power of air movement from the fan. Transmission unit which reduces RPM from the power source through a system of gears to the design speed for the fan to produce optimum air flow.

GIRT (See Horizontal Tie)

HANDRAIL

A horizontal or sloping rail placed along an access way or at the edge of a platform. Usually at 3′–6″ above the walkway or floor.

HEADER

In a counterflow tower, the main pipe carrying hot water to a series of laterals for distribution over the fill material. In a crossflow tower, the main pipe carrying hot water to each cell hot water distribution basin longitudinally. Crossover pipe carries the hot water laterally within the cell.

HEART WOOD

The wood extending from the pith to the sap wood, the cells of which no longer participate in the life processes of the tree. Heart wood may contain phenolic compounds, gums, resins, and other materials that usually make it darker and more decay resistant than sap wood.

HEAT EXCHANGE

A device for transferring heat from one substance to another. Heat transfer can be by direct contact, as in a cooling tower, or indirect, as in a shell and tube condensor.

HEAT LOAD

The amount of heat dissipated in a cooling tower measured in BTU's; is equal to the weight of water circulated per unit time multiplied by the cooling range. BTU per hr. = (GPM) (500) (HWT-CWT).

HORIZONTAL TIE

Framework member; a load-bearing horizontal connecting member in the tower framework.

HOT WATER TEMPERATURE

Temperature of circulating water entering the distribution system. Unit: °F.

HUMIDITY, ABSOLUTE

Amount of water vapor contained in the air at a given condition usually expressed in pounds of water per pounds of dry air.

HUMIDITY, RELATIVE

Ratio of water vapor pressure in the air to the saturated vapor pressure at the same temperature usually expressed as a percentage.

HYPERBOLIC TOWER

An evaporative type natural draft cooling tower of hyperbolic shape which depends on natural draft for air movement through the tower. Can be either crossflow or counterflow type of tower also Natural Draft Tower, or Atmospheric Tower.

INDUCED DRAFT

Unit with the air mover on top of the tower pulling the air up through the fill and out through the top, normally by an axial fan. In Mechanical Draft Towers, and Atmospheric Towers.

INLET CONNECTION (Inlet Flange)

Fitting to which the circulating water supply piping is connected to serve the tower distribution system.

JOISTS (or Beam)

Framework member; a horizontal support for live and dead loads such as fan decks, drift eliminators, elevated basins, piping. One of a series of parallel beams used to support floors or decking and supported in turn by larger beams.

LATENT HEAT

Heat required to change a liquid into a vapor without change in temperature or pressure.

LATERAL HEADER (Counterflow Tower)

A pipe or flume to distribute water from the distribution header to nozzles or other points of discharge to the filling area.

LONGITUDINAL

Always measured parallel to air intakes and perpendicular to air flow.

LOUVER

Baffles used for the purpose of changing the direction of air flow in a uniform parallel manner. Also utilized to prevent water droplets from splashing out of the tower on their descent through the structure.

LUMBER, BOARD

Wood that is nominally less than 2″ thick and two or more inches wide.

LUMBER, DRESSED SIZE

The dimensions of lumber after being surfaced with a plaining machine. The dressed size is usually ½″ to ¾″ less than a nominal or rough size.

LUMBER, NOMINAL SIZE

As applied to timber or lumber, the size by which it is known and sold in the market; which differs from the actual or dressed size.

LUMBER, ROUGH

Lumber which has not been dressed (surfaced) but has been sawed, edged, and trimmed.

LUMBER, SHIPLAPPED

Lumber that is edged dressed to make a lapped joint produced by placing one member partly over another and bonding the overlapped portions.

LUMBER, TIMBERS, BEAMS

Lumber that is nominally 5″ or more in least dimension.

MAKE-UP would have analysis of H_2O going back in

Replacement of water to the system lost by drift, evaporation, and purge, automatically controlled by a float valve.

MANIFOLD

The main header pipe in a Crossflow Tower.

MECHANICAL DRAFT

Design whereby airflow is constantly being generated by fans or rotor, either induced by the fans on top or forced draft by impellers located at bottom of air inlet.

MECHANICAL EQUIPMENT

Refers to air moving devices only. Induced draft normally consists of fan blades mounted on a hub attached to vertical shaft of gear reducer box with a flexible coupling fixed to the horizontal (or input shaft of gear box) and transmission shaft with another flexible coupling joining shaft to motor shaft. Forced draft towers normally are belt driven from motor pulling to fan sheave which is keyed to Rotor Fan (or squirrel cage) shaft. Some fan designs are direct coupled to motor.

MECHANICAL EQUIPMENT SUPPORT

Members which comprise the primary support for the fan drive assembly.

MERKEL'S THEORY

The total heat theory which has been almost universally adopted to calculations of tower performance. All heat transfer taking place at any position in the cooling tower is proportional to the difference between the total heat of the air at that point in the tower and a total of air saturated at the temperature of the water at that point in the tower.

MOTOR RATED HORSEPOWER

Horsepower rating inscribed on nameplate of the motor driving the fan.

NATURAL DRAFT WATER COOLING TOWER

One in which air movement is dependent upon the difference in density between the entering air and internal air. As the heat of the water is transferred to the air passing through the tower, the warmed air tends to rise and draw in fresh air at the base of the tower.

NET EFFECTIVE VOLUME

That portion of the total structural volume within which space the circulating water is in intimate contact with the airflow through the tower. Unit: cu.ft.

NOMINAL TOWER DIMENSIONS

Width and length measured from and to column center lines; height measured from top of basin curb to top of fan deck (counterflow design) or to top of distribution basin (crossflow design).

OBSTRUCTION LIGHTS (Natural Draft Tower)

Warning lights required by FAA regulations, placed on the outside of the hyperbolic shell.

OVERALL TOWER DIMENSIONS

(a) *width:* overall dimensions perpendicular to the tower's longitudinal axis; (b) *length*: overall dimension parallel to the air inlet louvers and the longitudinal axis; (c) total *height:* distance from basin curb to top of fan stack. Dimensions measured in feet.

PACKING—(See Filling)

PARTITION

An interior wall subdividing the tower into cells or into separate fan plenum areas.

PEDESTALS (Natural Draft Tower)

Used as a transition from the cross struts to the foundation.

PERFORMANCE CHARACTERISTIC, KAV/L

Relative efficiency factor for a given L/G ratio, volume, and cross section of a cooling tower, related to the design heat transfer capability of the fill, expressed by the formula KAV/L where "K" equals the coefficient of mass transfer in pounds per hour per square foot which is a unit of humidity potential. "A" equals total interfaced area per unit packed volume of tower in foot squared per foot cubed. "V" equals packed tower volume foot cubed per foot squared. "L" equals mass velocity of liquid in pounds per hour per foot squared of tower cross section. Liquid/Gas ratio relationship of water to air volume and water volume. This is a valuable tool used to rate the performance of a tower for heat transfer characteristics of a specific type of fill referred to as L/G or L over G.

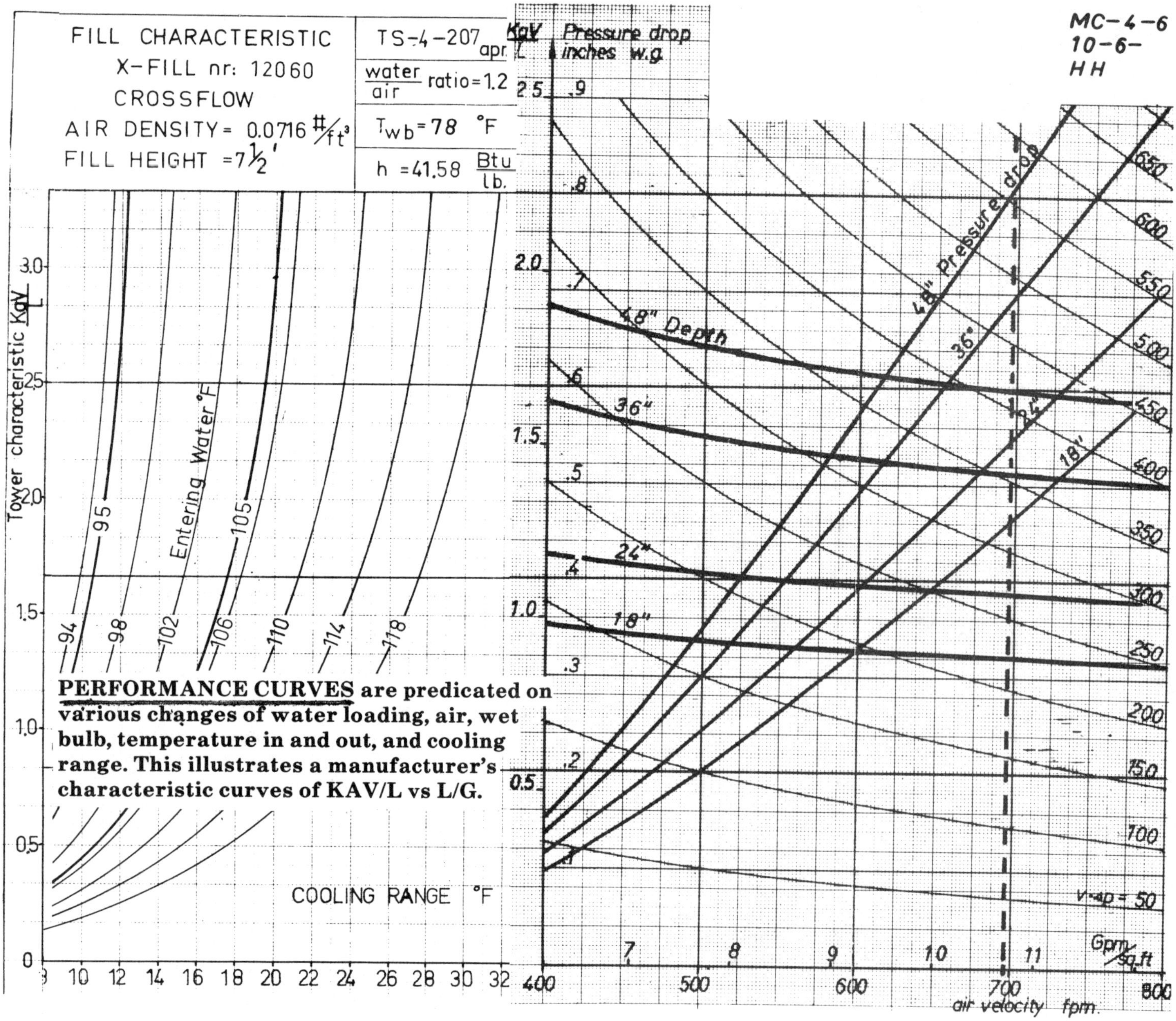

PERFORMANCE CURVES are predicated on various changes of water loading, air, wet bulb, temperature in and out, and cooling range. This illustrates a manufacturer's characteristic curves of KAV/L vs L/G.

PERFORMANCE CURVES

Performance of a cooling tower is rated in terms of water temperatures, approach, wet bulb temperature, range, static pressure, and air movement. The ratio of a cooling tower is established by developing a series of charts that relate to these variables.

pH OF WATER

The measure of purity, acidity, or alkalinity of water. 7.0 is arbitrarily selected as the neutral area where both the hydrogen and hydroxyl ions are present in exactly the same proportions so that the water is neutral. Lower than 7.0 is acid, higher is alkaline on a scale of 0 to 14.

PIER

In a wood tower an interior column support used to elevate column footings above the basic floor or foundation grade. Also known as Internal Pier.

PILASTER

An exterior column support, usually an integral part of the basin wall. Also known as External Pier.

PITCH ANGLE

Fan blade angle measured between the horizontal and upper side of the fan blade from leading edge to trailing edge. (discharge side)

PITCHING FAN BLADES

Angle of attack on fan blades of most modern cooling towers is adjustable. A bubble is initially set at the manufacturers recommended pitch angle at a designated place on each blade, the blade is turned to the angle and locked in place. This is done for all blades and the fan is rotated while the amperage is checked. Additional air volume or less air volume may be generated by the fans depending upon the pitch of the blade. It is important to know that the amperage of the system should remain approximately 10% below the motor plate.

PITOT TUBE

An instrument that operates on the principle of differential pressures. The primary use on cooling towers is the measurement of circulating water rate.

PLAN AREA

The cross section of the working area of the tower, measured in feet of length and width.

PLENUM

The enclosed space between the drift eliminators and the fan stack in induced draft towers, or the enclosed space between the fan and the filling in forced draft towers.

PLUME

Visible exhaust from a cooling tower.

POND

In a cooling pond heat is dissipated from the surface of a body of water by evaporation, radiation and convection. A spray pond was the original heat transfer cooling device. An improvement was the addition of an aeration spray extending the water surface by bringing it in contact with the air. The weaknesses of the cooling ponds are their dependence on the unpredictable prevailing winds, and excessive area of real estate required.

POWER FACTOR

The ratio of true power (watts) to the apparent power, as indicated by the product of amps × volts.

PRESSURE DROP

Static pressure or amount of resistance of air movement through the tower caused by obstructions such as air louver inlets, fill configuration, tower supports, spray system piping, drift eliminators, static and mechanical equipment components, fan cylinder resistance and construction.

PSYCHROMETER

An instrument used primarily to measure the wet bulb temperatures. Either a sling or a mechanically aspirated type of psychrometer is acceptable provided the instrument is properly shielded from radiation and the air across the wick is limited to approximately 1,000 ft./min.

PUMPING HEAD

Energy required to raise the water to the distribution elevation and overcome friction losses through pipe, valves, fittings and nozzles. It is expressed in feet of liquid the pump must move and equals the totals of friction loss, static head, plus pressure drop through the distribution system.

PURGE

Removing the concentrated impurities from evaporation or discharging water from the system to prevent precipitation of solids. (Blowdown).

RADIATION

Heat loss caused by a body warmer than surrounding air giving off its latent heat by flowing thermally "downhill" from the hot body to the cooler area.

RANGE

The numerical difference between the water temperature entering the cooling tower at the distribution system and the cold water temperature leaving the sump of the cooling tower.

RECIRCULATION

A phenomenon where the hot exhaust air is forced downward and back into the cooling tower mixing with and lowering the cool fresh inlet air, usually due to design or placement problems. This raises the wet bulb temperature of the entering air above that of the ambient air and greatly reduces tower performances.

REDISTRIBUTION BASIN

An elevated basin installed between the hot and cold water basins in a crossflow tower to maintain uniform water distribution throughout the entire height of the fill.

RISER

Piping which connects the circulating water supply line from the level at the base of the tower or the supply header to the tower inlet connection.

SAFETY

A cooling tower is dangerous. Fan blades can and have killed. A power lockout is mandatory at the fan location. If unavailable—a watch should be established at the power source if it cannot be padlocked out. Initial fan start-up must be visually cleared prior to energizing, Fan, rail, ladder, and louver guards are to be maintained. A Buddy system should be utilized for interior repairs. Primary mandate is simple—use common sense.

SAFETY HANDRAIL

Railing around top of tower, platforms and stairways; usually composed of a top handrail, kneerail and toeboard (depending on location).

SAPWOOD

The wood of pale color near the outside of the log. Under most conditions the sapwood is more susceptible to decay than heart wood.

SENSIBLE HEAT

The "heat you can feel" the heat required to change the temperature of the air or water. Hot water gives up a portion of its heat to the colder air by the heat "flowing downhill" from hot to cold.

SHELL DIAMETER (Natural Draft Tower)

Diameter of the shell at the top of the curb, measured from inside of cross struts to outside of cross struts.

SHELL HEIGHT (Natural Draft Tower)

Dimension from top of curb to top of the hyperbolic shell.

SINGLE FLOW WATER COOLING

A Crossflow Tower having a fill section on one side of the plenum chamber only.

SOFFIT (Natural Draft Tower)

The underside of the ring beam, forming the bottom part of the hyperbolic shell. Generally it is thick at the base and is the first portion of the shell above the cross struts.

SOFT ROT

Decay developing under very wet conditions in the outer wood layers, caused by cellulose destroying microfungi that attack the secondary cell walls and not the inter cellular layer.

SPEED REDUCER

A device to change speed of the driver to the speed desired for the fan.

SPLASH BAR

Horizontal component of a fill deck in a tower which constitutes the principal splash surface.

SPLASHER

Used in a gravity distribution system to receive water from a downspout and effect uniform spreading of the water over the wetted area of the tower.

SPRAY FILLED WATER COOLING TOWER

A tower which has no fill. Water to air contact depends entirely on the break-up of the water by means of pressure spray nozzles.

SPRAY NOZZLE

Device used in a distribution system to break up the flow of the circulating water into droplets and effect uniform spreading of the water over the wetted area of the tower.

STATIC PRESSURE DROP

Reduction of air movement through the tower due to the resistance of internal components such as air intake louvers, fill packing, water distribution system, internal supporting beams, drift eliminators, fan stack configuration.

STANDARD AIR

Has a density of 0.075 lb. per cu. ft., which is air at 68°F DBT, 50 percent RH and 29.92 in. HG (and substantially equivalent to 70°F dry air).

SUCTION SCREEN

Screen over water sump or pump suction pit on tower cold water basin, prevents debris from entering pump and piping system.

SUMP

Depressed section of the cold water collecting basin where the cooled water is returned to the heat source, be it the condensors, compressors, or process water equipment.

SUPPLY HEADER

Portion of the water supply system from which riser or inlet connection receives the circulating water flow.

THERMAL CAPACITY

The number of gallons per minute (GPM) a cooling tower will handle for a specified range, wet bulb temperature, and approach. Also simply known as CAPACITY.

THROAT DIAMETER (Natural Draft Tower)

Diameter of the shell at its narrowest point.

TOWER DIMENSIONS

(a) Width: overall dimension perpendicular to the tower longitudinal axis; (b) Length: overall dimension parallel to the air inlet louvers and the longitudinal axis; (c) Total height: distance from top of basin curb to top of fan stack.

TRANSVERSE

Always measured perpendicular to air intakes.

UNDERFLOW WATER COOLING TOWER

A crossflow tower with a forced draft fan located in a horizontal plane at the bottom of the plenum chamber. Air from the plenum travels horizontally through fill sections and discharges vertically at each end of the tower.

VELOCITY REGAIN STACK

A venturi shaped fan cylinder. The increased in performance obtained on this type of cylinder is due to the reduction in total head above the fan. The attendant decrease in total head differential across the fan results in an increase in air rate at constant horse power or a decrease in horse power at constant air rate.

VIBRATION CUT-OUT SWITCH

Safety switch used to de-energize the driver circuit should the mechanical equipment assembly vibrate excessively.

WATER COOLING TOWER

An enclosed, steady flow device for cooling water by evaporation and sensible heat exchange through direct contact with air.

WATER LOAD

Circulation rate of water over the tower, irrespective of temperature range.

WATER LOADING

Circulating water flow expressed in gallons per minute (GPM) per square foot of effective horizontal wetted area of the tower. Unit: GPM/ft.^{2}.

WET BULB

The temperature of saturated air. At 100% humidity the wet bulb temperature equals the dry bulb reading. The lower the humidity, the drier the air and the greater the difference in reading. The lower the wet bulb, the more work a cooling tower can do. A cooling tower cannot cool the water below the wet bulb temperature of the inlet air.

WET BULB TEMPERATURE (WBT)

Temperature indicated by a psychrometer. Also known as the thermodynamic wet bulb temperature or the temperature of adiabatic saturation. Unit: °F.

WET/DRY TOWER

A wet (evaporative) cooling tower in combination with a dry (non-evaporative) heat exchanger system, used to reduce or abate cooling tower fog during cold weather by modifying the tower exhaust air condition.

WIND LOAD

The load applied to a structure by a wind blowing against its surface. Wind load is expressed in pounds per square foot. Unit: lbs./ft.^{2}.

WINTERIZATION

Modification of one or more cells to operate under freezing ambient temperatures. Consists of installing steam or electrical heating coils in basin, additional water distribution system below the fill packing to by-pass going through the decks, together with necessary valving, and fan reversal switch, if required, to keep heat in tower if water cannot by-pass fill to help melt ice formation.

APPENDIX 2

METRIFICATION

talking about cooling towers... you may find the following main conversion factors useful

This textbook was compiled in part from articles written by the author, together with other pertinent information required to obtain a continuity of information.

Much of this treatise was prepared prior to the acceptance of metrics in the U.S. and is written in english units. In preparation for when the Cooling Tower Industry will standardize on "Le Systeme International d' Unites" commonly called International System of Units (SI Units), the following conversion tables are listed which will be helpful in transposing cooling tower measurements and terms from one category to the other.

CONVERSION AND EQUIVALENCY TABLE

	MEASURE GIVEN	MEASURE REQUIRED	MULTIPLIER
A.	Atmosphere	inch Hg	29,921
	Atmosphère	inch Water (15 °C)	407,14
	Atmosphère	pound (Av)/square inch	14,6959
	Atmosphère	kg/cm²	1,0332
B.	barrel (Petroleum)	m³	0,15897
	B.T.U.	kcal	0,251996
	B.T.U.	kgm	107,580
	B.T.U./sec	CV	1,434
	B.T.U./pound (Av)	kcal/kg	0,55554
	B.T.U./cu.ft	kcal/m³	8,899
	B.T.U./square inch	kcal/m²	390,57
C.	°C	°F °C	1,8 + 32
	cm	inch	0,3937
	cm²	square inch	0,155
	cm²	square foot	0,001076
	cm³	cubic inch	0,061023
	cm³	cubic foot	0,000035314
	cubic foot	l	28,3168
	cubic foot	m³	0,02831677
	cubic foot/long ton	m³/t	0,027869
	cubic foot/net ton	m³/t	0,031215
	cubic foot/pound (Av)	m³/kg	0,062428
	cubic inch	cm³	16,3872
	cubic yard	m³	0,76455
	CV	B.T.U./sec	0,6972
	CV	foot pound (Av)/sec	542,50
	CV	HP	0,9863
	CV	kWatt	0,7355
F.	°F	°C (°F — 32)	0,5555
	fluid ounce	cm³	29,573
	foot	m	0,3047997
	foot pound (Av)	Joule	1,3551
	foot pound (Av)	kgm	0,13825
	foot pound (Av)/sec	CV	0,0018434
	foot pound (Av)/sec	Watt	1,3551
G.	g	grain	15,43236
	g	ounce (Av)	0,035274
	g	pound (Av)	0,0022046
	g	pound (Troy)	0,002679
	g/cm²	pound/sq. inch	0,01422
	g/cm³	pound (Av)/cubic foot	62,42
	g/l	pound (Av)/gallon (G.-B.)	0,010017
	g/l	pound (Av)/gallon (U.S.A.)	0,008345
	g/m³	grain/cubic foot	0,43701
	gallon (G.-B.) Imperial	l	4,5459
	gallon (U.S.A.)	l	3,7854
	gallon (G.-B.)	cubic foot	0,16045
	gallon (U.S.A.)	cubic foot	0,13368
	gallon (G.-B.)/long ton	l/t	4,4718
	gallon (U.S.A.)/net ton	l/t	4,1727
	gallon (G.-B.)/square yard	l/m²	5,4340
	gallon (U.S.A.)/square yard	l/m²	4,5273
H.	HP	kcal/sec	0,1782
	HP	kWatt	0,7455
	HP	kgm/sec	76,042
	HP	CV	1,0139
	hpz	kg/cm²	1,0197
	hpz	pound (Av) sq. inch	14,5037
	hundredweight (cwt) (G.-B.)	kg	50,8024
I.	inch	mm	25,4
	inch Hg	Atmosphère	0,03342
	inch Hg	pound (Av) sq. inch	0,49115
	inch Hg	kg/cm²	0,034534
	inch water	mm/Hg	1,867
	inch water	Atmosphère	0,0024583
	inch water	kg/cm²	0,002541
J.	Joule	foot pound	0,7376
K.	kcal	B.T.U.	3,9683
	kcal	foot pound	3,086
	kcal/sec	HP	5,6142
	kcal/kg	B.T.U./pound (Av)	1,8001
	kcal/kg	therm/long ton	0,040303
	kcal/kg	therm/net ton	0,035985
	kcal/m²	B.T.U./square inch	0,0025604
	kcal/m³	B.T.U./cubic foot	0,11237
	kcal/t	B.T.U./long ton	4,0323
	kcal/t	B.T.U./net ton	3,6001
	kg	long ton (G.-B.)	0,0009842
	kg	net ton (U.S.A.)	0,0011023
	kg	ounce (Av)	35,274
	kg	pound (Av) ou lb	2,20462
	kg/cm²	inch Hg	28,958
	kg/cm²	inch water	393,55
	kg/cm²	hpz	0,9806
	kg/cm²	pound (Av)/square inch	14,2234
	kg/cm	pound (Av)/inch	5,5997
	kgm	B.T.U.	0,0092956
	kgm	foot pound (Av)	7,2330
	kgm/sec	HP	0,013151
	kg/m	pound (Av)/foot	0,67197
	kg/m²	pound (Av)/square foot	0,20482
	kg/m³	pound (Av)/cubic foot	0,0623
	kg/m³	pound (Av)/gallon (G.-B.)	0,008345
	kg/t	pound (Av)/long ton	2,2400
	kg/t	pound (Av)/net ton	2,0
	km	mile (nautical)	0,53961
	km	mile (statute)	0,62137
	knot	km	1,8533
	kWatt	HP	1,3418
	kWatt	CV	1,3596
	kWh	kcal	860,57
	kWh	B.T.U.	3.412,76
	kWh	foot pound (Av)	2.655.180
L.	l	bushel	0,028378
	l	cubic inch	61,0250
	l	cubic foot	0,035315
	l	gallon (G.-B.)	0,2201
	l	gallon (U.S.A.)	0,26418
	l	pint (G.-B.)	1,7621
	l	pint (U.S.A.)	2,1134
	l	quart (U.S.A.)	1,0567
	l/t	gallon (G.-B.)/long ton	0,22363
	l/t	gallon (U.S.A.)/net ton	0,23965
	lb (pound Avoir du poids)	kg	0,4535924
	lb (pound Troy)	kg	0,37324
	long ton (G.-B.)	kg	1.016,047
M.	m	inch	39,370
	m	foot	3,280843
	m	yard	1,0936
	m²	square inch	1.550,00
	m²	square foot	10,7639
	m²	square yard	1,19399
	m²	acre	0,00024711
	m³	barrel 42 gallons	6,290495
	m³	cubic foot	35,314759
	m³	cubic inch	61.023,9034
	m³	cubic yard	1,3080
	m³	gallon (G.-B.)	220,10
	m³	gallon (U.S.A.)	264,18
	m³	pint	2.113,4
	m³/kg	cubic foot/pound (Av)	16,0185
	m³/t	cubic foot/long ton	35,883

	m³/t	cubic foot/net ton	32,036
	mile (nautical)	km	1,8533
	mile (statute)	km	1,60935
	mm	inch	0,039370
	mm Hg	inch water	0,5357
	mm Hg	pound (Av) square inch	0,0193368
	metric ton	short ton	1,1023
	metric ton	long ton	0,9842
N.	net ton (short ton) U.S.A.	kg	907,185
P.	pound (Avoir du poids)	kg	0,4535924
	pound (Troy)	kg	0,37324
	pound (Av)/cubic foot	g/cm³ (kg/l)	0,016019
	pound (Av)/cubic inch	kg/cm³	0,027680
	pound (Av)/gallon (G.-B.)	g/l	99,832
	pound (Av)/gallon (U.S.A.)	g/l	119,83
	pound (Av)/inch	kg/cm	0,17858
	pound (Av)/long ton	kg/t	0,44643
	pound (Av)/net ton	kg/t	0,5000
	pound (Av)/square foot	kg/m²	4,8824
	pound (Av)/square inch	kg/cm²	0,070307
	pound (Av)/square inch	Atmosphère	0,068046
	pound (Av)/square inch	mm Hg	51,7149
Q.	quart (U.S.A.)	l	0,9464
S.	short ton (net ton) U.S.A.	kg	907,185
	square foot	m²	0,092903
	square inch	cm²	6,45163
	square yard	m²	0,83613
	square mile	km²	2,58989
T.	t	long ton	0,98421
	t	net ton (U.S.A.)	1,10231
	therm (100.000 B.T.U.)	kcal	25.200,0
	therm/long ton	kcal/kg	24,802
	therm/net ton (U.S.A.)	kcal/kg	27,790
Y.	yard	m	0,91440

°C		°F	°C		°F	°C		°F	°C		°F	°C		°F	°C		°F
-18.9	-2	28.4	2.2	36	96.8	23.3	74	165.2	44.4	112	233.6	65.6	150	302.0	86.7	188	370.4
18.3	1	30.2	2.8	37	98.6	23.9	75	167.0	45.0	113	235.4	66.1	151	303.8	87.2	189	372.2
-17.8	0	32.0	3.3	38	100.4	24.4	76	168.8	45.6	114	237.2	66.7	152	305.6	87.8	190	374.0
-17.2	1	33.8	3.9	39	102.2	25.0	77	170.6	46.1	115	239.0	67.2	153	307.4	88.3	191	375.8
16.7	2	35.6	4.4	40	104.0	25.6	78	172.4	46.7	116	240.8	67.8	154	309.2	88.9	192	377.6
16.1	3	37.4	5.0	41	105.8	26.1	79	174.2	47.2	117	242.6	68.3	155	311.0	89.4	193	379.4
15.6	4	39.2	5.6	42	107.6	26.7	80	176.0	47.8	118	244.4	68.9	156	312.8	90.0	194	381.2
15.0	5	41.0	6.1	43	109.4	27.2	81	177.8	48.3	119	246.2	69.4	157	314.6	90.6	195	383.0
14.4	6	42.8	6.7	44	111.2	27.8	82	179.6	48.9	120	248.0	70.0	158	316.4	91.1	196	384.8
13.9	7	44.6	7.2	45	113.0	28.3	83	181.4	49.4	121	249.8	70.6	159	318.2	91.7	197	386.6
13.3	8	46.4	7.8	46	114.8	28.9	84	183.2	50.0	122	251.6	71.1	160	320.0	92.2	198	388.4
12.8	9	48.2	8.3	47	116.6	29.4	85	185.0	50.6	123	253.4	71.7	161	321.8	92.8	199	390.2
12.2	10	50.0	8.9	48	118.4	30.0	86	186.8	51.1	124	255.2	72.2	162	323.6	93.3	200	392.0
-11.7	11	51.8	9.4	49	120.2	30.6	87	188.6	51.7	125	257.0	72.8	163	325.4	93.9	201	393.8
11.1	12	53.6	10.0	50	122.0	31.1	88	190.4	52.2	126	258.8	73.3	164	327.2	94.4	202	395.6
10.6	13	55.4	10.6	51	123.8	31.7	89	192.2	52.8	127	260.6	73.9	165	329.0	95.0	203	397.4
10.0	14	57.2	11.1	52	125.6	32.2	90	194.0	53.3	128	262.4	74.4	166	330.8	95.6	204	399.2
9.4	15	59.0	11.7	53	127.4	32.8	91	195.8	53.9	129	264.2	75.0	167	332.6	96.1	205	401.0
8.9	16	60.8	12.2	54	129.2	33.3	92	197.6	54.4	130	266.0	75.6	168	334.4	96.7	206	402.8
8.3	17	62.6	12.8	55	131.0	33.9	93	199.4	55.0	131	267.8	76.1	169	336.2	97.2	207	404.6
7.8	18	64.4	13.3	56	132.8	34.4	94	201.2	55.6	132	269.6	76.7	170	338.0	97.8	208	406.4
7.2	19	66.2	13.9	57	134.6	35.0	95	203.0	56.1	133	271.4	77.2	171	339.8	98.3	209	408.2
6.7	20	68.0	14.4	58	136.4	35.6	96	204.8	56.7	134	273.2	77.8	172	341.6	98.9	210	410.0
6.1	21	69.8	15.0	59	138.2	36.1	97	206.6	57.2	135	275.0	78.3	173	343.4	99.4	211	411.8
5.6	22	71.6	15.6	60	140.0	36.7	98	208.4	57.8	136	276.8	78.9	174	345.2	100.0	212	413.6
5.0	23	73.4	16.1	61	141.8	37.2	99	210.2	58.3	137	278.6	79.4	175	347.0	100.6	213	415.4
4.4	24	75.2	16.7	62	143.6	37.8	100	212.0	58.9	138	280.4	80.0	176	348.8	101.1	214	417.2
3.9	25	77.0	17.2	63	145.4	38.3	101	213.8	59.4	139	282.2	80.6	177	350.6	101.7	215	419.0
-3.3	26	78.8	17.8	64	147.2	38.9	102	215.6	60.0	140	284.0	81.1	178	352.4	102.2	216	420.8
2.8	27	80.6	18.3	65	149.0	39.4	103	217.4	60.6	141	285.8	81.7	179	354.2	102.8	217	422.6
-2.2	28	82.4	18.9	66	150.8	40.0	104	219.2	61.1	142	287.6	82.2	180	356.0	103.3	218	424.4
1.7	29	84.2	19.4	67	152.6	40.6	105	221.0	61.7	143	289.4	82.8	181	357.8	103.9	219	426.2
1.1	30	86.0	20.0	68	154.4	41.1	106	222.8	62.2	144	291.2	83.3	182	359.6	104.4	220	428.0
0.6	31	87.8	20.6	69	156.2	41.7	107	224.6	62.8	145	293.0	83.9	183	361.4	105.0	221	429.8
0.0	32	89.6	21.1	70	158.0	42.2	108	226.4	63.3	146	294.8	84.4	184	363.2	105.6	222	431.6
0.6	33	91.4	21.7	71	159.8	42.8	109	228.2	63.9	147	296.6	85.0	185	365.0	106.1	223	433.4
1.1	34	93.2	22.2	72	161.6	43.3	110	230.0	64.4	148	298.4	85.6	186	366.8	106.7	224	435.2
1.7	35	95.0	22.8	73	163.4	43.9	111	231.8	65.0	149	300.2	86.1	187	363.6	107.2	225	437.0

APPENDIX 3

PSYCHOMETRIC ANALYSIS

PSYCHROMETRIC ANALYSIS

THERMAL PERFORMANCE OF A COOLING TOWER IS FFECTED BY THE ENTERING WET BULB TEMPERATURE.

AIR ENTERS AT AMBIENT POINT A - ABSORBS HEAT AND MOISTURE FROM THE WATER - EXITING AT POINT B IN A SATURATED CONDITION.

THE AMOUNT OF HEAT TRANSFERRED FROM THE WATER TO THE AIR IS PROPORTIONAL TO THE DIFFERENCE IN ENTHALPY OF THE AIR BETWEEN THE ENTERING AND LEAVING CONDITIONS $[h_B - h_A]$.

VECTOR AB MAY BE SEPARATED INTO COMPONENTS AC, WHICH REPRESENTS SENSIBLE HEAT EXCHANGE, AND COMPONENT CB, WHICH REPRESENTS LATENT HEAT EXCHANGE.

IF THE ENTERING AIR CONDITION IS NOW POINT D AT THE SAME WET BULB TEMPERATURE, BUT A DIFFERENT DRY BULB TEMPERATURE, THE TOTAL HEAT TRANSFER REMAINS THE SAME BUT THE SENSIBLE AND LATENT COMPONENTS HAVE CHANGED FOR THE SAME WATER LOAD. THE AMOUNT OF

EVAPORATION DEPENDS UPON THE AMOUNT TO SENSIBLE HEATING OR COOLING OF THE AIR.

THE RATIO OF LATENT TO SENSIBLE HEAT IS IMPORTANT IN ANALYZING THE WATER USAGE OF A COOLING TOWER. MASS TRANSFER [EVAPORATION] OCCURS ONLY IN THE LATENT PORTION OF THE HEAT TRANSFER PROCESS AND IS PROPORTIONAL TO THE CHANGE IN SPECIFIC HUMIDITY.

SINCE THE ENTERING AIR DRY BULB TEMPERATURE OR RELATIVE HUMIDITY EFFECTS THE LATENT TO SENSIBLE HEAT TRANSFER RATIO, IT ALSO EFFECTS THE RATE OF EVAPORATION. THE RATE OF EVAPORATION IS LESS IN AB [W_B - W_A] THEN IN DB [W_B - W_D] BECAUSE THE LATENT HEAT TRANSFER [MASS TRANSFER] REPRESENTS A SMALLER PORTION OF THE TOTAL.

THE EVAPORATION RATE AT TYPICAL DESIGN CONDITIONS IS APPROXIMATELY 1% OF THE WATER CIRCULATION RATE FOR EACH 10°F OF WATER TEMPERATURE RANGE.

IN ADDITION TO WATER LOSS OF EVAPORATION, LOSS CAN ALSO OCCUR DUE TO LIQUID CARRY OVER INTO THE DISCHARGE AIR STREAM, WINDAGE LOSSES, LEAKS, MALFUNCTIONING LOUVERS, AND BLOWDOWN TO MAINTAIN ACCEPTABLE WATER QUALITY.

EXCHANGE OF SENSIBLE HEAT PRODUCES ABOUT 25% OF THE HEAT TRANSFER EXCHANGE OF LATENT HEAT OF EVAPORATION DUE TO THE EVAPORATION OF A PORTION OF THE WATER USUALLY APPROXIMATELY 1% OF THE WATER FLOW FOR EACH 10°F OF TEMPERATURE RANGE PRODUCES ABOUT 75% OF THE HEAT TRANSFER IN A COOLING TOWER.

PLUMES (FOG)

THE WARM AIR DISCHARGE FROM THE COOLING TOWER IS ESSENTIALLY SATURATED UNDER CERTAIN OPERATING CONDITIONS THE AMBIENT AIR SURROUNDING THE TOWER IS UNABLE TO ABSORB ALL OF THE MOISTURE IN THE DISCHARGE STREAM AND THE EXCESS CONDENSES AS VISABLE FOG. FOGGING CAN BE PREDICTED BY DRAWING A STRAIGHT LINE ON THE PSYCHROMETRIC CHART FROM THE TOWER ENTERING CONDITIONS TO THE POINT REPRESENTING THE DISCHARGE CONDITIONS. THE GREATER THE AREA OF INTERSECTION TO THE LEFT OF THE SATURATION CURVE, THE MORE INTENSE THE PLUME.

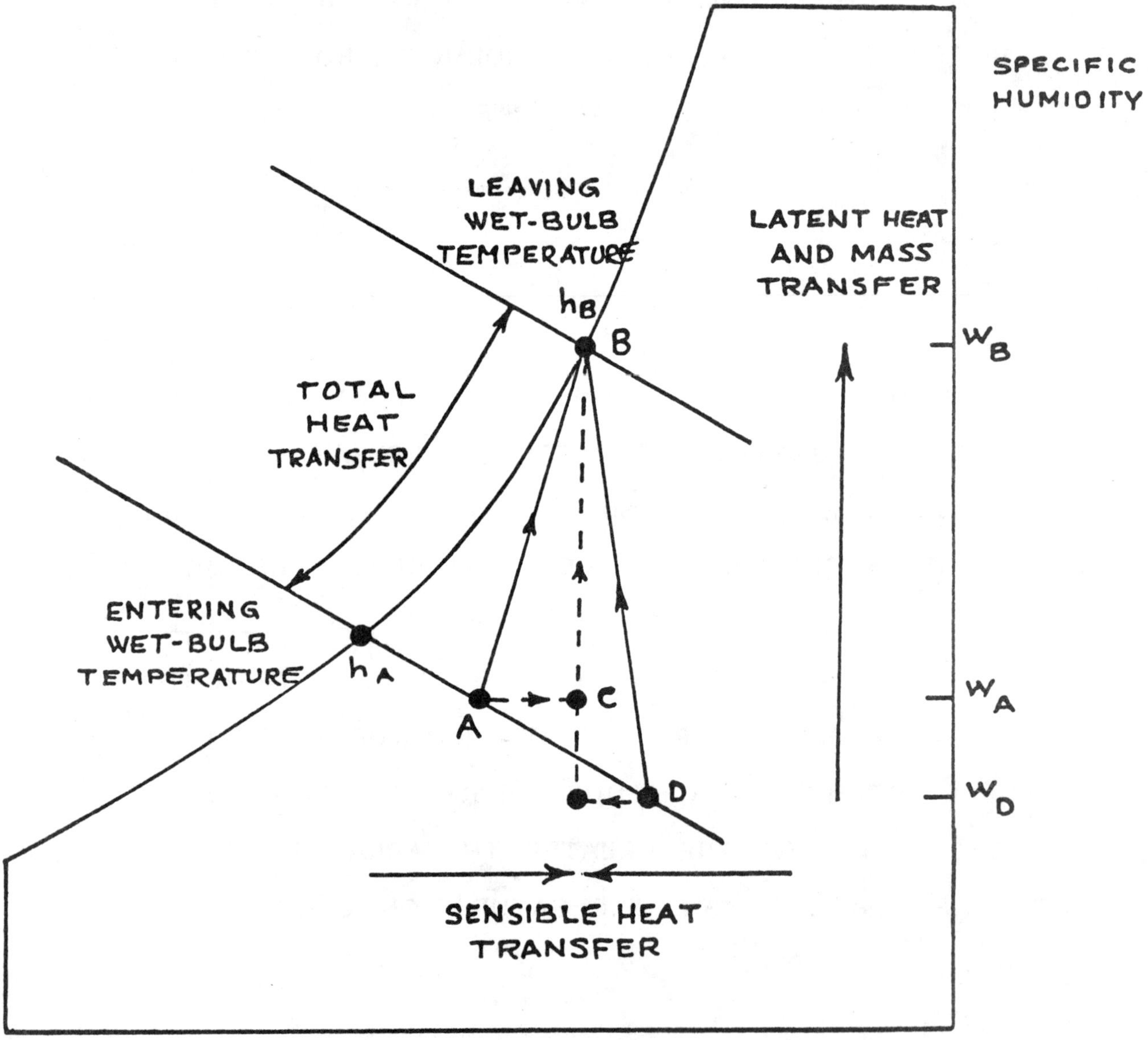

PSYCHROMETRIC ANALYSIS
OF AIR PASSING THRU
COOLING TOWER

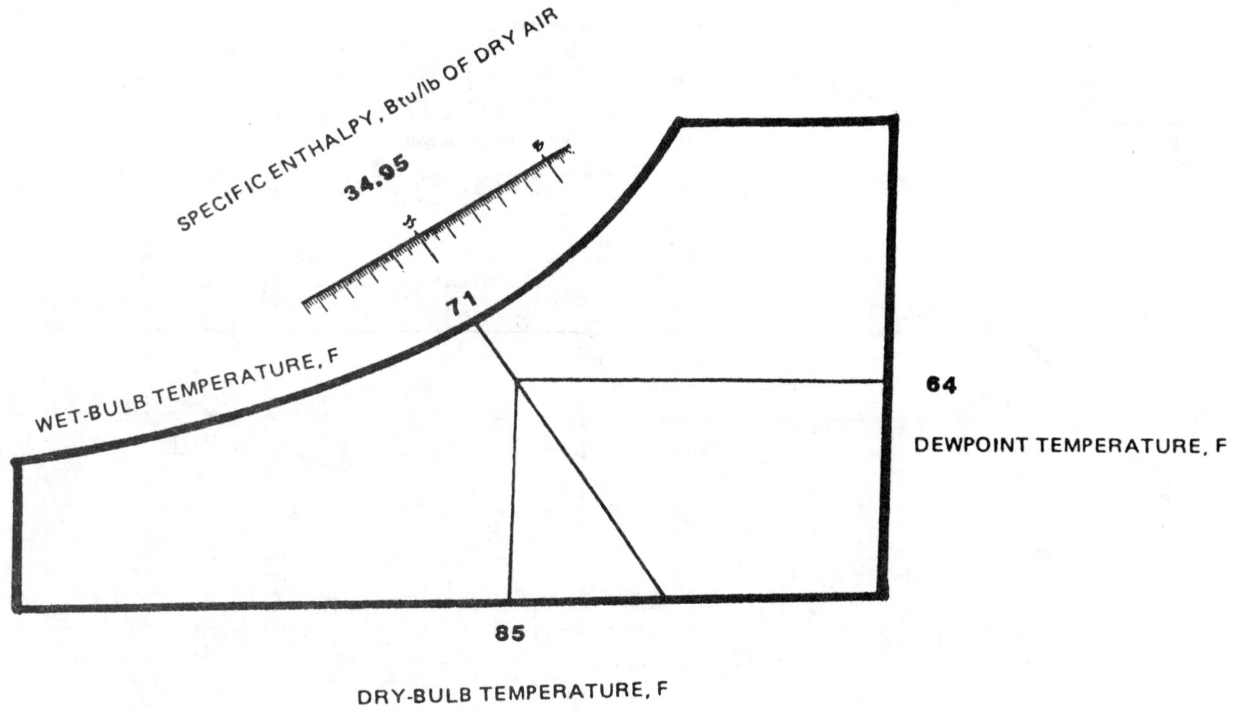

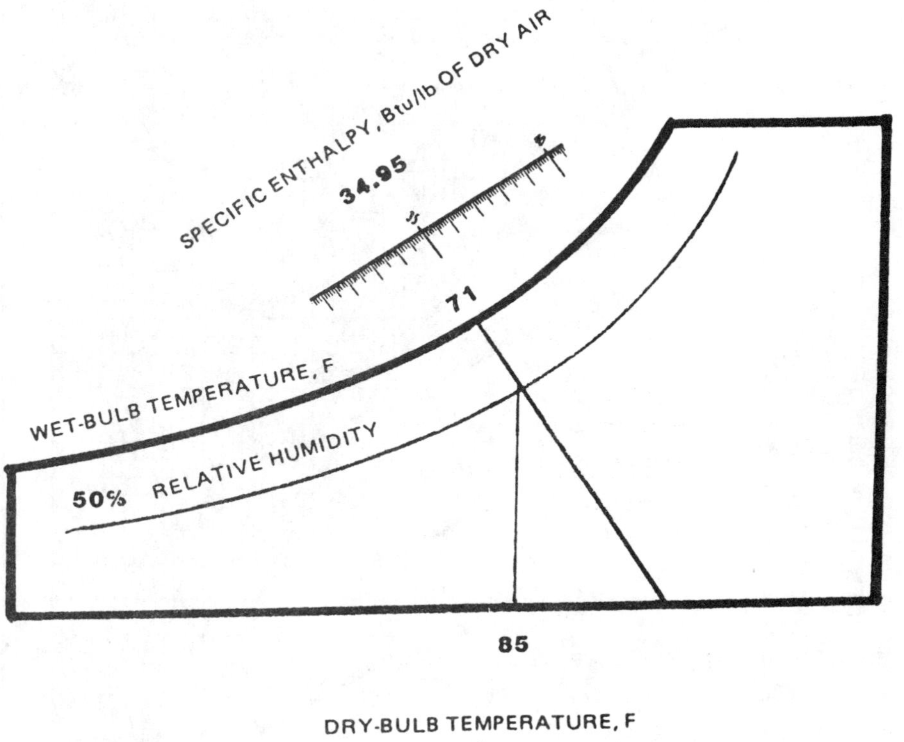

VARIABLES DETERMINING WET BULB

PSYCHROMETRIC CHART

1 DRY-BULB TEMPERATURE, F
2 WET-BULB TEMPERATURE, F
3 HUMIDITY RATIO, gr/lb OF DRY AIR
4 SPECIFIC ENTHALPY, Btu/lb OF DRY AIR
5 SPECIFIC VOLUME, cu ft/lb OF DRY AIR
6 DEWPOINT TEMPERATURE, F
7 RELATIVE HUMIDITY, PERCENT
8 VAPOR PRESSURE, in.Hg

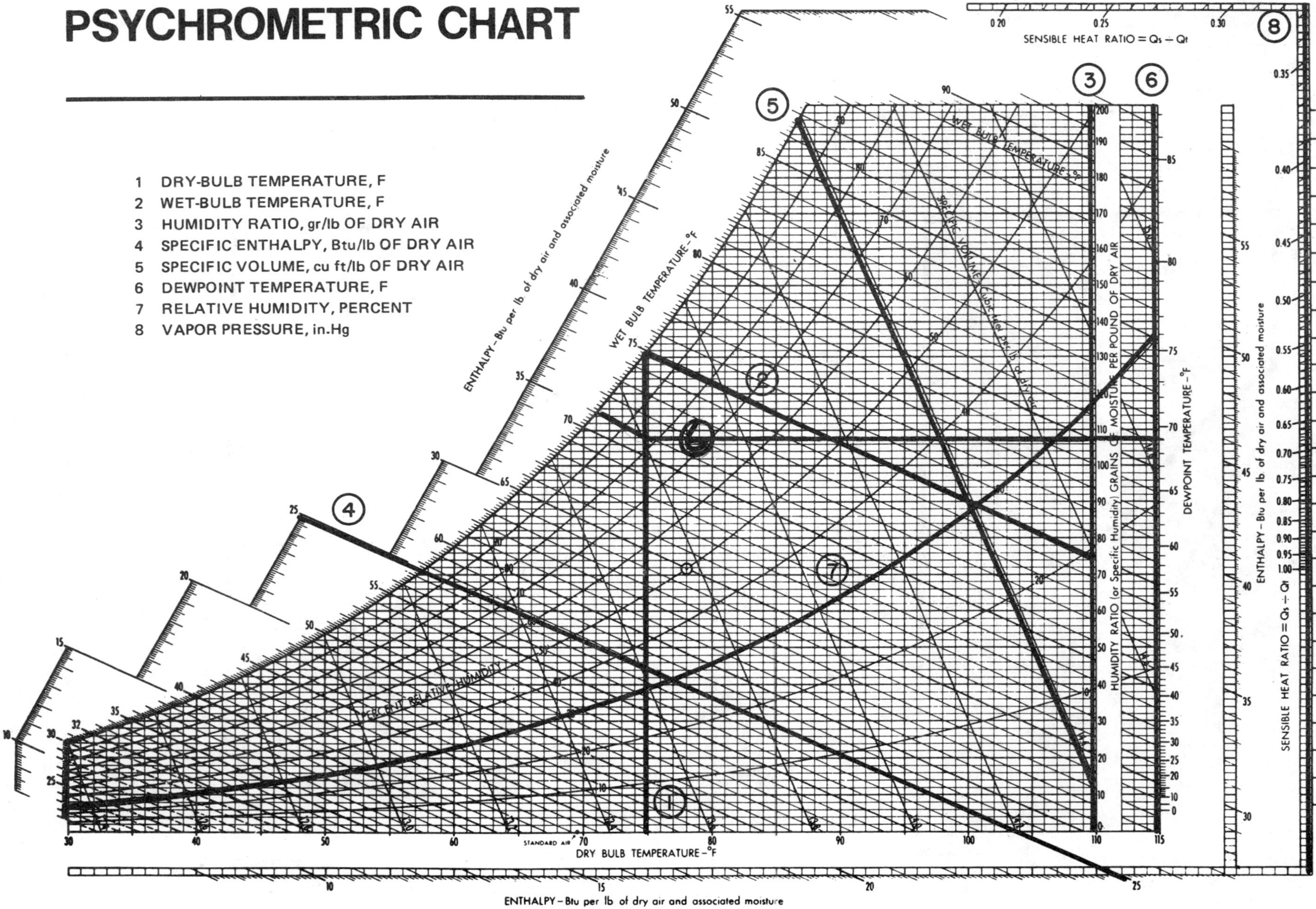

To further assist in using this "new language", two psychrometric charts are illustrated, one at 760 mm Hg and the other at 29.92 inches of mercury (760 mm × 0.3937 = 29.92 in Hg).

ENTHALPIC AIR DIAGRAM AT 760 mm Hg

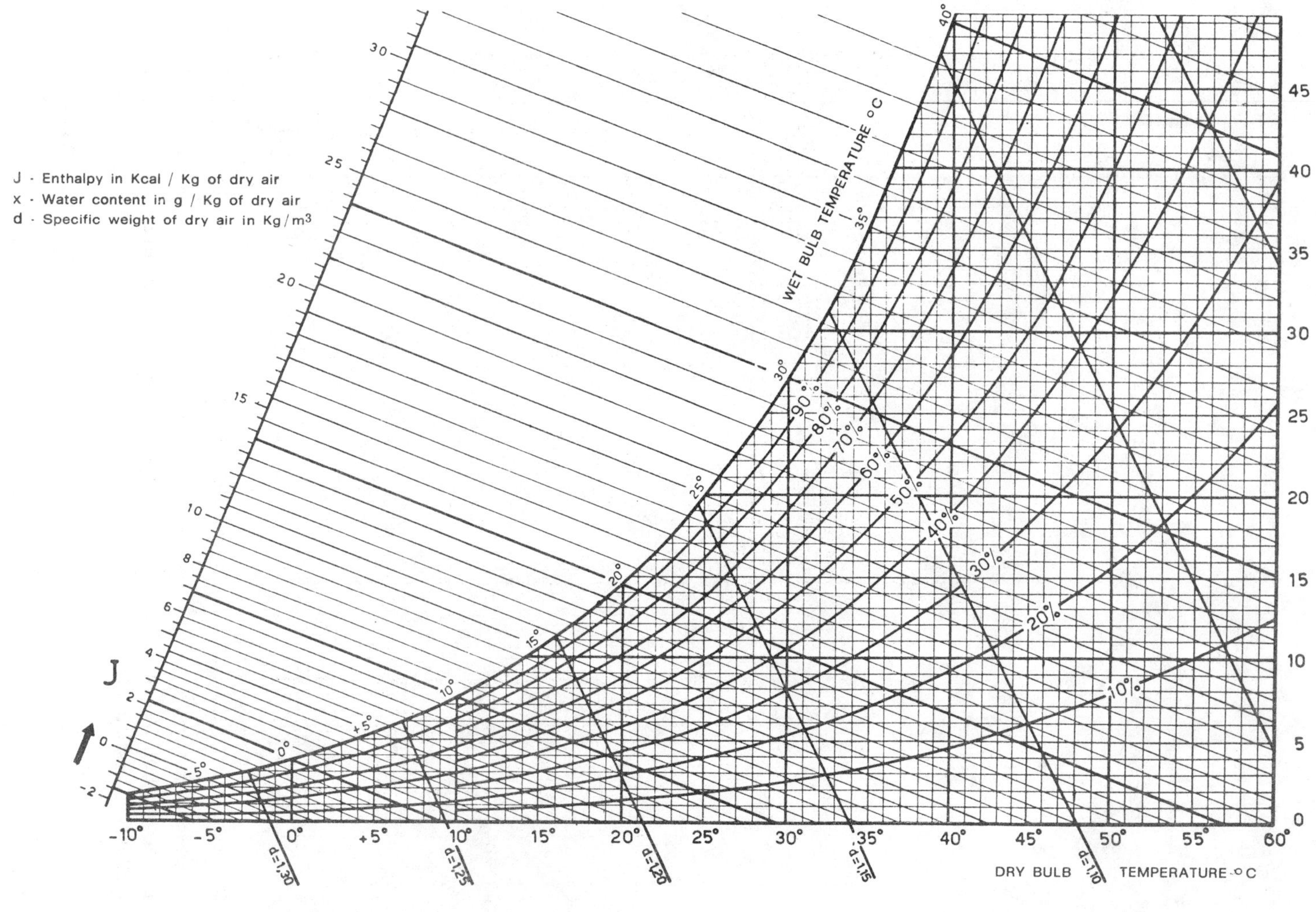

CHART USE

DATA GIVEN:

- Ambient air dry bulb temperature: 25 °C
- Relative humidity: 50 %

DATA FROM CHART:

- Wet bulb temperature: 18 °C
- Enthalpy of air: 12 kcal/kg.
- Air density: 1,17 Kg/m³
- Water content: g in each Kg. of dry air: 10 g/Kg.